PROPAGATORS IN QUANTUM CHEMISTRY

A Series of Monographs

THEORETICAL CHEMISTRY

Consulting Editors

D. P. CRAIG

Research School of Chemistry, Institute of Advanced Studies, Australian National University, Canberra, Australia

R. McWEENY

Department of Chemistry, University of Sheffield, Sheffield, England

---

Volume 1 T. E. PEACOCK: Electronic Properties of Aromatic and Heterocyclic Molecules

Volume 2 R. McWEENY and B. T. SUTCLIFFE: Methods of Molecular Quantum Mechanics

Volume 3 J. LINDERBERG and Y. ÖHRN: Propagators in Quantum Chemistry

Forthcoming Volumes

E. STEWART: The Spectra of Small Polyatomic Molecules
A. C. HURLEY: The Electronic Theory of Small Molecules
D. P. SANTRY and H. W. KROTO: Semi-empirical Molecular Orbital Theory
T. THIRUNAMACHANDRAN: Theory of Valence in the Heavier Elements
I. G. ROSS: Spectra of Aromatic and Heterocyclic Molecules
C. D. H. CHISHOLM: Group Theoretical Techniques in Quantum Chemistry
S. M. BLINDER: Time-dependent Perturbation Theory

PROPAGATORS IN QUANTUM CHEMISTRY

JAN LINDERBERG

Department of Chemistry, Aarhus University
Aarhus, Denmark

and

YNGVE ÖHRN

Departments of Chemistry and Physics
University of Florida, Gainsville, Florida, U.S.A.

1973

ACADEMIC PRESS: LONDON AND NEW YORK

A Subsidiary of Harcourt Brace Jovanovich, Publishers

ACADEMIC PRESS INC. (LONDON) LTD.
24/28 Oval Road
London NW1

United States Edition published by
ACADEMIC PRESS INC.
111 Fifth Avenue
New York, New York 10003

Library of Congress Catalog Card Number: 72-12274
ISBN: 0 12 450350 0

PRINTED IN GREAT BRITAIN BY
Adlard & Son Ltd., Bartholomew Press, Dorking

PREFACE

This book has emerged from a desire on the part of the authors to present a molecular approach to many-electron theory and the awareness of a growing interest for such a text. The necessary stimulus for initiating this project was provided by the organizers of a number of advanced study institutes by inviting us to lecture on many-body topics. We are grateful to the Hungarian Academy of Sciences, through Professor R. Gáspár, The Battelle Memorial Research Institute, through Professor C. W. Kern, to Uppsala University and the University of Florida, through Professor P. O. Löwdin, for financial support. The Spring term of 1968 was spent by Jan Linderberg at the Department of Chemistry, University of Florida, when the manuscript was started, and Yngve Öhrn spent the academic year of 1970–71 at the Department of Chemistry, Aarhus University, when the bulk of the manuscript was prepared. We are indebted to both these institutions for providing the opportunity for this close collaboration.

Our interest in the many-electron problem and the use of second quantization was implanted by our teacher, Professor P. O. Löwdin, and we appreciate his continued interest in this work. Guidance in the use of propagators and steady encouragement by Professor S. Lundqvist is gratefully acknowledged.

Some of the numerical results that are presented have been obtained with the assistance of S. F. Abdulnur, E. Dalgaard, P. G. C. Kaijser, P. Lindner, and P. W. Thulstrup. Their able work is highly appreciated.

We thank Lisegrete Blach for preparing the typescript with skill and patience.

Aarhus and Gainsville
April, 1973

JAN LINDERBERG
YNGVE ÖHRN

CONTENTS

CHAPTER 1

The Purpose of Propagators in Quantum Chemistry

Chemical experiments give information of the nature of matter only in relation to a theoretical interpretation. Conversely it can be maintained that a theoretical development is relevant for chemistry only to the extent that it facilitates these interpretations. We advance the opinion that the propagator is an extremely flexible and rich concept, which pervades all of theoretical chemistry and links experiment with theory.

Stationary states cannot, strictly speaking, be observed as such through experiments, since every observation involves a probing of the system with external "perturbations". The response to such a probing of the molecular system and its influence back on the probe are directly involved in measuring processes. A theoretical analysis focusing on the calculation of observable quantities leads immediately to the study of propagators as the basic entity. This view was pioneered by Richard P. Feynman in his path integral approach to quantum mechanics, and propagators are now generally used tools in physics.

The success of the propagator approach in physics is connected with the solution of some complicated many-body problems. Noteworthy is the treatment of collective motion in an electron gas and its relation to exchange and correlation effects in the theory of electronic band structure of metals. Such progress spurred the interest in the development of similar methods for molecular electronic systems. Spectacular advances have yet to be made in quantum chemistry, but it is the aim of this text to show that propagators and second quantization can be a resource for the theoretical development in chemistry although the full machinery of many-particle physics is not invoked. To this end we have not stressed new methods but emphasized well-known results as obtained with the language of second quantization and propagators. Although it is seldom directly used in the text, the reader should be aware of the possibility the propagators offer for a direct connection between microscopic and macroscopic descriptions.

The text is designed for the graduate student or research worker, who is already familiar with atomic and molecular quantum theory and with elementary statistical physics. The choice of topics reflects to a large extent the personal prejudices of the authors, but it is hoped that such a wide range of material is covered that the general usefulness of the approach is clearly demonstrated. Some important developments in the many-electron theory of atoms and molecules have been omitted for the reason that they were judged to fall outside the scope of this book.

Problems have been added at the end of the chapters in order to assist the reader in checking his ability to handle the necessary algebra. They also give some illustrative examples of further applications of the methods.

Notes and Bibliography

The reference level for the knowledge of quantum mechanics expected of the reader is that of L. I. Schiff's "Quantum Mechanics" (McGraw-Hill, 1955, 2nd edition). A useful survey of molecular quantum theory is offered by "The Quantum Theory of Molecular Electronic Structure", a lecture note and reprint volume by R. G. Parr (W. A. Benjamin, 1963). L. Hedin and S. Lundqvist have given a comprehensive review of many-body theory as applied to the electronic band structure problem in solids, in "Solid State Physics" (Editors Seitz, Turnbull, and Ehrenreich, Academic Press) **23**, 1 (1969). More general questions of many-particle theory are dealt with in "The Many-Body Problem in Quantum Mechanics" (Cambridge, 1967) by N. H. March, W. H. Young and S. Sampanthar.

CHAPTER 2

Differential Equations and their Green's Functions

Solutions to the differential equations met with in quantum mechanics are often conveniently and elegantly discussed in terms of the corresponding integral equations. This technique, which employs the concept of Green's function is commonly used in scattering theory but is rarely found in textbooks on quantum chemistry. We construct some simple examples of Green's functions to ordinary differential equations in order to demonstrate the importance of boundary conditions.

The linear, ordinary differential equation of the form

$$[E+\tfrac{1}{2}d^2/dx^2-V(x)]\,\psi(x)=0, \tag{2.1}$$

has the general solution

$$\psi(x)=Au(x)+Bv(x), \tag{2.2}$$

where $u(x)$ and $v(x)$ are two linearly independent solutions to Equation (2.1) and A and B are constants. One requires generally in quantum mechanical problems that $\psi(x)$ should satisfy boundary conditions, such as being zero at the end points of an interval (a, b). Satisfactory solutions to Equation (2.1) can then be obtained only for certain eigenvalues $E=\epsilon_n$.

An inhomogenous equation similar to Equation (2.1),

$$[E+\tfrac{1}{2}d^2/dx^2-V(x)]\,\psi(x)=r(x), \tag{2.3}$$

admits a solution with the appropriate boundary conditions fulfilled. Let us assume that the previously discussed linearly independent solutions $u(x)$ and $v(x)$ of Equation (2.1) have been found and determined such that

$$u(a)=v(b)=0. \tag{2.4}$$

We know that their wronskian, $W=u(x)v'(x)-u'(x)v(x)$, is constant since

$$dW/dx=u(x)v''(x)-u''(x)v(x)=0, \tag{2.5}$$

by virtue of Equation (2.1).

A solution to Equation (2.3) can now be constructed by first multiplying by $2u(x)$ to obtain

$$\frac{d}{dx}\,[u\psi'-u'\psi]=2ur, \tag{2.6}$$

and by $2v(x)$ to obtain

$$\frac{d}{dx}[v\psi' - v'\psi] = 2vr. \tag{2.7}$$

Secondly, both these equations are integrated, which gives

$$u(x)\psi'(x) - u'(x)\psi(x) = 2\int_a^x u(y)r(y)dy \tag{2.8}$$

and

$$v'(x)\psi(x) - v(x)\psi'(x) = 2\int_x^b v(y)r(y)dy, \tag{2.9}$$

when the conditions on $\psi(x)$ are invoked. The derivative $\psi'(x)$ can be eliminated from Equations (2.8) and (2.9) and we get

$$\psi(x) = (2/W)\left[v(x)\int_a^x u(y)r(y)dy + u(x)\int_x^b v(y)r(y)dy\right]. \tag{2.10}$$

This expression can be written in terms of the Green's function $G(x, x'; E)$ as

$$\psi(x) = \int_a^b G(x, x'; E)r(x')dx'. \tag{2.11}$$

A comparison between Equations (2.10) and (2.11) gives the definition

$$G(x, x'; E) = (2/W)u(x_<)v(x_>), \tag{2.12}$$

where $x_<$ and $x_>$ are the smaller and the larger respectively of x and x'.

It is evident that the Green's function (2.12) is a solution to Equation (2.3) for the particular choice of source function $r(x) = \delta(x - x')$. Thus it holds that

$$[E + \tfrac{1}{2}d^2/dx^2 - V(x)]\, G(x, x'; E) = \delta(x - x'). \tag{2.13}$$

The treatment of Equations (2.3) to (2.9) is valid for any value of the parameter E, but the step that leads to Equation (2.10) needs further consideration when E equals one of the eigenvalues ϵ_n to the homogeneous Equation (2.1). In such a case both $u(x)$ and $v(x)$ are proportional to the corresponding eigenfunction $\psi_n(x)$ and no longer linearly independent and thus the wronskian equals zero. There will still be a solution to Equation (2.3) if $r(x)$ is orthogonal to $\psi_n(x)$,

$$\int_a^b \psi_n(x)r(x)dx = 0. \tag{2.14}$$

It follows that Equation (2.8) and (2.9) are identical.

On an interval where $\psi_n(x)$ is nodeless we obtain from Equation (2.8) that

$$\frac{d}{dx}\left[\frac{\psi(x)}{\psi_n(x)}\right] = 2[\psi_n(x)]^{-2}\int_a^x \psi_n(y)r(y)dy. \tag{2.15}$$

The further integration of Equation (2.15) is simplified considerably by the introduction of a function $Q(x)$ such that

$$dQ(x)/dx=[\psi_n(x)]^{-2}, \tag{2.16}$$

when x is different from the nodal points of $\psi_n(x)$.

This allows us to rewrite Equation (2.15) as

$$\frac{d}{dx}\left[\frac{\psi(x)}{\psi_n(x)}-2Q(x)\int_a^x \psi_n(y)r(y)dy\right]=-2Q(x)\psi_n(x)r(x), \tag{2.17}$$

where the right hand side will be shown to be integrable over the interval (a, b). Thus we have a solution

$$\psi_+(x)=2\psi_n(x)\int_a^x \psi_n(y)r(y)\,[Q(x)-Q(y)]dy, \tag{2.18}$$

or by an equivalent procedure

$$\psi_-(x)=2\psi_n(x)\int_x^b \psi_n(y)r(y)\,[Q(y)-Q(x)]dy. \tag{2.19}$$

Both these equations can be written as integral equations that are similar in form to Equation (2.11) but where the kernels are denoted $K_+(x, x'; \epsilon_n)$ and $K_-(x, x'; \epsilon_n)$ respectively.

We have

$$K_+(x, x'; \epsilon_n)=\begin{cases}0, & x<x'\\ 2\psi_n(x)\psi_n(x')\,[Q(x)-Q(x')], & x>x'\end{cases} \tag{2.20}$$

and

$$K_-(x, x'; \epsilon_n)=\begin{cases}2\psi_n(x)\psi_n(x')\,[Q(x')-Q(x)], & x<x'\\ 0, & x>x'\end{cases} \tag{2.21}$$

from which we obtain

$$\psi_\pm(x)=\int_a^b K_\pm(x, x'; \epsilon_n)r(x')dx'. \tag{2.22}$$

A third kernel, the arithmetic mean of K_+ and K_-, is more symmetric and gives a third result,

$$K(x, x'; \epsilon_n)=\tfrac{1}{2}K_+(x, x'; \epsilon_n)+\tfrac{1}{2}K_-(x, x'; \epsilon_n), \tag{2.23}$$

$$\psi(x)=\int_a^b K(x, x'; \epsilon_n)r(x')dx'. \tag{2.24}$$

The orthogonality condition (2.14) was necessary to insure the compatibility of Equations (2.8) and (2.9) for the case of $E=\epsilon_n$. It is readily seen that

$$\psi_\pm(x)=\psi(x)\mp\psi_n(x)\int_a^b \psi_n(x')Q(x')r(x')dx', \tag{2.25}$$

and that a general solution of Equation (2.3) for the case of $E=\epsilon_n$ will differ from $\psi(x)$ as given in Equation (2.24) by an arbitrary multiple of $\psi_n(x)$. There is correspondingly a family of kernels that can be written

$$K_{AB}(x, x'; \epsilon_n)=K(x, x'; \epsilon_n)+A(x)\psi_n(x')+\psi_n(x)B(x'). \tag{2.26}$$

Thus K_+ corresponds to the choice

$$A(x) = -B(x) = \psi_n(x)Q(x), \tag{2.27}$$

and K_- obtains when

$$B(x) = -A(x) = \psi_n(x)Q(x). \tag{2.28}$$

We propose to show that a kernel K_{AB} exists such that

$$\lim_{E\to\epsilon_n} [G(x, x'; E) - \psi_n(x)\,[E-\epsilon_n]^{-1}\psi_n(x')] = K_{AB}(x, x'; \epsilon_n). \tag{2.29}$$

It is evident from the form of G that this kernel is symmetric with respect to interchange of x and x' and that consequently $A(x) = B(x)$.

The eigenvalue equation for $\psi_n(x)$ can be written as

$$[E + \tfrac{1}{2}d^2/dx^2 - V(x)]\,\psi_n(x) = (E-\epsilon_n)\,\psi_n(x), \tag{2.30}$$

which according to Equation (2.11) is equivalent to the integral equation

$$\psi_n(x) = (E-\epsilon_n)\int_a^b G(x, x'; E)\,\psi_n(x')dx'. \tag{2.31}$$

From Equations (2.11) and (2.31) it follows that

$$\int_a^b \psi_n(x)\psi(x)dx = (E-\epsilon_n)^{-1}\int_a^b \psi_n(x)r(x)dx, \tag{2.32}$$

and that for values of E different from ϵ_n, $\psi(x)$ will be orthogonal to $\psi_n(x)$ simultaneously with $r(x)$. We conclude that the limiting kernel of Equation (2.29) is such that

$$\int_a^b K_{AA}(x, x'; \epsilon_n)\psi_n(x')dx' = 0. \tag{2.33}$$

The result for the function $A(x)$ from Equation (2.33) is conveniently expressed in terms of the auxiliary functions,

$$N(x) = \int_a^x |\,\psi_n(x')\,|^2\,dx', \tag{2.34}$$

and

$$M(x) = \int_a^x N'(x')Q(x')dx'. \tag{2.35}$$

We obtain the expression

$$\begin{aligned} A(x)N(b) + \psi_n(x)\int_a^b A(x')\psi_n(x')dx' &= -\int_a^b K(x, x'; \epsilon_n)\psi_n(x')dx' \\ &= -\psi_n(x)\,[2Q(x)N(x) - Q(x)N(b) - 2M(x) + M(b)], \end{aligned} \tag{2.36}$$

which yields the integral

$$N(b)\int_a^b A(x)\psi_n(x)dx = 2\int_a^b M(x)N'(x)dx - M(b)N(b), \tag{2.37}$$

after some partial integrations.

There remains to be discussed the possible singularities of the function $Q(x)$ defined by Equation (2.16) and the right hand side of Equation (2.17). The eigenfunction $\psi_n(x)$ has only simple zeros in the interior of the interval (a, b). Let the nodal points be $[x_i]$, such that

$$a=x_0<x_1<x_2 \ldots x_n<x_{n+1}=b, \tag{2.38}$$

and

$$\psi_n(x_i)=0, \quad i=0, 1, 2, \ldots n, n+1. \tag{2.39}$$

It holds, however, that the derivative is nonvanishing at these points, except possibly at the end points,

$$\psi_n'(x_i)\neq 0, \quad i=1, 2, 3, \ldots n. \tag{2.40}$$

Obviously one can integrate Equation (2.16) to define a function $Q(x)$ on each open interval (x_{i-1}, x_i). A connection between different intervals can be obtained from the identity

$$\frac{d}{dx}\left[Q(x)+\frac{1}{\psi_n'(x)\psi_n(x)}\right]=2[\epsilon_n-V(x)]\,[\psi_n'(x)]^{-2}, \tag{2.41}$$

where the right hand side is regular at the nodal points. These considerations show that we can define an integral $Q(x)$ to Equation (2.16) on the open interval (a, b) such that its only singularities are simple poles at the nodal points of $\psi_n(x)$. The function on the right hand side is thus integrable for reasonable source functions $r(x)$. Special care needs to be exercised at the end points of the interval where the derivative $\psi_n'(x)$ may vanish simultaneously with $\psi_n(x)$.

Two distinct cases need to be examined for possible integration problems at the end points of the interval. The first and simplest arises when the source function $r(x)$ is such that the integrals (2.18) and (2.19) exist. Then the rearrangements were permissible that led to Equations (2.22) and (2.24). The second case, namely when the integration of the right hand side of Equation (2.17) cannot be extended to the end points of the interval gives no problem if the intermediate steps are omitted and direct use is made of the limiting kernel of Equation (2.29). The contributions to K_{AA} from the divergent functions of the form $\psi_n(x)Q(x)$ will cancel and the only condition on the function $r(x)$ will be that the integral (2.14) is well defined.

Hall's Minimum Principle

A reformulation of the eigenvalue problem (2.1) was carried out by Hall, by means of Green's functions, for the case of a potential $V(x)$, which is negative on the interval (a, b). The procedure applies to bound state eigenvalues, $E<0$. We consider

$$[E+\tfrac{1}{2}d^2/dx^2]\,\psi(x)=V(x)\psi(x), \tag{2.42}$$

which is of the form (2.3), and choose $\kappa>0$ such that $E=-\frac{1}{2}\kappa^2$.

Then it follows that

$$u(x)=\sinh \kappa(x-a), \tag{2.43}$$

$$v(x)=\sinh \kappa(x-b), \tag{2.44}$$

and

$$W=\kappa \sinh \kappa(b-a). \tag{2.45}$$

The Green's function from Equation (2.12) is employed to give

$$\psi(x)=\int_a^b G(x,x';E)V(x')\psi(x')dx'. \tag{2.46}$$

Hall considered the functional,

$$J(\chi)=\int_a^b dx\int_a^b dx'\chi(x)V(x)G(x,x';E)V(x')\chi(x')\Big/\int_a^b dx\,\chi^2(x)V(x), \tag{2.47}$$

and concluded that it is stationary and equal to unity for $\chi(x)=\psi(x)$. Both the numerator and the denominator are negative since $V(x)$ is negative on the interval and the numerator can be rewritten as

$$-2\int_a^b dx\left[\int_a^x dx'u(x')V(x')\chi(x')/u(x)\right]^2.$$

The minimum property of the functional $J(\chi)$ ascertains that if, for a chosen value κ, a function χ can be found such that $J(\chi)=1$, then $E=-\frac{1}{2}\kappa^2$ is an upper bound to the smallest eigenvalue of Equation (2.1).

A proof of the bounding property of E with respect to the ground state eigenvalue of Equation (2.42) follows from the ordinary variational principle,

$$E(\chi)=\int_a^b \chi(x)\,[-\tfrac{1}{2}d^2/dx^2+V(x)]\,\chi(x)dx\Big/\int_a^b dx\,\chi^2(x)\geq\epsilon_0. \tag{2.48}$$

We introduce the function $\tilde{\chi}$,

$$\tilde{\chi}(x)=\int_a^b dx'G(x,x';E)V(x')\chi(x'), \tag{2.49}$$

and apply the variational principle (2.48) to the trial function $N[\chi(x)+\lambda\tilde{\chi}(x)]$. The constant N is to be chosen such that the norm of the trial function equals unity. Calculation of the functional $E(\chi+\lambda\tilde{\chi})$ yields

$$\epsilon_0\leq E+N^2[E(\chi)-E+\lambda^2\int_a^b dxV(x)\,[\tilde{\chi}(x)-\chi(x)]^2], \tag{2.50}$$

under the assumption that E is chosen such that $J(\chi)=1$.

The term proportional to λ^2 is negative and $E(\chi)>E$ since

$$E(\chi)=E+\tfrac{1}{2}\int_a^b dx[\kappa^2|\tilde{\chi}-\chi|^2+|d(\tilde{\chi}-\chi)/dx|^2], \tag{2.51}$$

and there is thus a real value λ for which the parentheses in relation (2.50) vanishes.

The conclusion of the previous argument is the relations

$$E(\chi) \geq E \geq \epsilon_0, \tag{2.52}$$

which show that Hall's minimum principle gives a better upper bound than the ordinary variational principle for a given trial function $\chi(x)$. More general functions $\chi(x)$ are admitted for the functional $J(\chi)$ than for $E(\chi)$ but the procedure is limited to negative potentials only.

Problem 1

Demonstrate that Hall's functional $J(\chi)$ for the Yukawa potential $V(x) = -(\mu/x)\exp(-\alpha x)$ and $\chi(x) = x\exp(-\beta x)$ equals

$$J = \mu(\alpha+2\beta)^2/(\alpha+\beta)(\alpha+\beta+\kappa)^2$$

when the interval is $(0, \infty)$. Solve the equations

$$J = 1 \text{ and } dJ/d\beta = 0$$

for β and κ when $\alpha = 5$ and $\mu = 6{\cdot}5$. Compare the energy value $-\frac{1}{2}\kappa^2$ with the accurate value $-2{\cdot}216$ obtained by Hylleraas and Risberg, and the value $-1{\cdot}973$ calculated from $E(\chi)$. [$\beta = 5{\cdot}1115$, $\kappa = 2{\cdot}0938$, $E = -2{\cdot}1920$.]

Problem 2

Calculate the wave function to first order in the perturbation from the spin–orbit coupling term for an electron in a hydrogen-like 3p-state in a Coulomb field. Convenient units are used such that $E = \epsilon_{3\text{p}} = -\frac{1}{2}$, $V(x) = x^{-2} - 3x^{-1}$, $\psi_{3\text{p}} = Nx^2(2-x)\,\mathrm{e}^{-x}$, and $r(x) = A(3x^{-3} - 1)\psi_{3\text{p}}$ in the notations of Equation (2.3). Check the orthogonality between $\psi_{3\text{p}}$ and $r(x)$ on the interval $(0, \infty)$. [Hint: Use direct integration of Equation (2.15).]

Notes and Bibliography

A comprehensive account of differential equations and their Green's functions is given by E. L. Ince (1956) in "Ordinary Differential Equations", Dover, New York. The applications to perturbation theory are reviewed by J. O. Hirschfelder, W. Byers Brown, and S. T. Epstein (1964) in "Advances in Quantum Chemistry" (P. O. Löwdin, ed.), Vol. 1, pp. 255–374. Academic Press, New York. G. G. Hall (1967) in *Chem. Phys. Letters* **1**, 495–496, presented the variational method based on the functional $J(\chi)$ and G. G. Hall, J. Hyslop, and D. Rees (1969) in *Intern. J. Quantum Chem.* **3**, 195–204, demonstrate the bounding property. The eigenvalue for the Yukawa potential quoted in the first problem to this chapter was obtained from E. A. Hylleraas and V. Risberg (1941) *Avh. Norske Vidensk.-Akad. Mat.-naturv. Kl.* No. 3.

CHAPTER 3

Schrödinger's Equations and their Green's Functions

The quantum mechanical description of electronic motion, as deduced by Schrödinger, is based on the concept of a wave function $\psi(\bar{r}t)$ as the probability amplitude for finding the particle at the space point $\bar{r}$ at time t. Schrödinger's wave equation tells us the time development of the amplitude when the initial conditions are given. We investigate these problems in a slightly more general context. The space coordinates $\bar{r}$ are generalized to space–spin coordinates ξ and we prefer to work with the wave function $\psi(\xi t)$ as expanded in a set of spin orbitals $u_s(\xi)$, which is assumed to be an orthonormal set. Thus we have

$$\psi(\xi t) = \sum_s u_s(\xi) a_s(t), \tag{3.1}$$

and the matrix form of Schrödinger's equation ($\hbar = 1$)

$$i da_s(t)/dt = \sum_r h_{sr} a_r(t). \tag{3.2}$$

This necessitates the study of systems of linear differential equations and the associated Green's functions. A preliminary account is given in this chapter of the physical content of these Green's functions and the connected propagator concept. Emphasis is placed on manipulations of analytical and algebraic character that will remain useful also in more complicated contexts than the one studied here.

The hamiltonian matrix $\mathbf{h} = [h_{sk}]$ is self adjoint and has a set of orthonormal eigenvectors $\mathbf{x} = [x_{sk}]$ such that the following relations are valid:

$$\sum_r h_{sr} x_{rk} = \epsilon_k x_{sk}, \tag{3.3}$$

$$\sum_s x^{\dagger}_{ks} h_{sr} = \epsilon_k x^{\dagger}_{kr}, \tag{3.4}$$

$$\sum_s x^{\dagger}_{ks} x_{sl} = \delta_{kl}, \tag{3.5}$$

$$\sum_k x_{sk} x^{\dagger}_{kr} = \delta_{sr}, \tag{3.6}$$

$$x^{\dagger}_{ks} = x^{*}_{sk}. \tag{3.7}$$

The matrix $\mathbf{x}$ is thus unitary and transforms the hamiltonian $\mathbf{h}$ to diagonal form.

Similarly there is a spectral resolution of **h** in the form

$$h_{sr} = \sum_k x_{sk} \epsilon_k x^\dagger_{kr}. \tag{3.8}$$

It follows from Equation (3.2) that

$$i \frac{d}{dt} [\sum_s x^\dagger_{ks} a_s(t)] = \epsilon_k [\sum_r x^\dagger_{kr} a_r(t)], \tag{3.9}$$

and that consequently

$$a_s(t) = \sum_k x_{sk} \exp[-i\epsilon_k(t-t')] \sum_r x^\dagger_{kr} a_r(t'). \tag{3.10}$$

The probability interpretation of quantum mechanics defines the probability for an electron to be "observed" in the spin orbital s at time t if it was known to be in the spin orbital r at time t' as

$$P_{sr}(t, t') = |\sum_k x_{sk} \exp[-i\epsilon_k(t-t')] x^\dagger_{kr}|^2. \tag{3.11}$$

A scattering experiment would be performed such that an electron is entering a system at time t' in the spin orbital r and is detected with the probability $P_{sr}(t, t')$ at time t in the spin orbital s.

We consider next the case of several electrons already present in the system but we disregard their interaction presently. The system of electrons is assumed to be in its ground state, which means that N electrons occupy the N lowest energy levels ϵ_k. An injected electron has access only to the unoccupied levels and in the probability amplitude we must now exclude contributions from occupied levels due to the Pauli principle. Furthermore we should realize that an electron which entered the system at time t' cannot be observed at times prior to t' and we define a probability $P'_{sr}(t, t')$ with these conditions included;

$$P'_{sr}(t, t') = \begin{cases} |\sum_k x_{sk}(1-f_k) \exp[-i\epsilon_k(t-t')] x^\dagger_{kr}|^2, & t > t' \\ 0, & t < t' \end{cases} \tag{3.12}$$

$$f_k = \begin{cases} 1, & k \text{ occupied,} \\ 0, & k \text{ unoccupied.} \end{cases} \tag{3.13}$$

An electron may very well be observed in the spin orbital s at an earlier time than t' if it is one of the electrons already present in the N-particle system. The previously described experiment was designed to observe an electron in spin orbital s and this is formally similar to ejecting an electron from this spin orbital. When an electron has been removed at time t from an occupied spin orbital s, we are left with a system of $N-1$ particles or equivalently a hole in the ground state configuration. The hole propagates so that there will be a probability $P''_{sr}(t, t')$ to observe it, by means of adding an electron, in the spin orbital r at the later time t':

$$P''_{sr}(t,t')=\begin{cases}0, & t>t' \\ |\sum_k x_{sk}f_k \exp[-i\epsilon_k(t-t')]x^\dagger_{kr}|^2, & t<t'.\end{cases} \tag{3.14}$$

The two probabilities $P'_{sr}(t,t')$ and $P''_{sr}(t,t')$ are derived from a common probability amplitude, the *propagator* $G_{sr}(t,t')$, which is defined as

$$G_{sr}(t,t')=\begin{cases}-i\sum_k x_{sk}(1-f_k)\exp[-i\epsilon_k(t-t')]x^\dagger_{kr}, & t>t' \\ i\sum_k x_{sk}f_k \exp[-i\epsilon_k(t-t')]x^\dagger_{kr}, & t<t'.\end{cases} \tag{3.15}$$

Thus we have that

$$|G_{sr}(t,t')|^2=\begin{cases}P'_{sr}(t,t'), & t>t' \\ P''_{rs}(t,t'), & t<t'.\end{cases} \tag{3.16}$$

It is discontinuous at $t=t'$ for $s=r$ which is shown by

$$\lim_{\delta\to+0}[G_{sr}(t,t-\delta)-G_{sr}(t,t+\delta)]=-i\delta_{sr}. \tag{3.17}$$

The propagator is also the Green's function to the Schrödinger equation (3.2):

$$i\partial G_{sr}(t,t')/\partial t-\sum_q h_{sq}G_{qr}(t,t')=\delta_{sr}\delta(t-t'), \tag{3.18}$$

where the δ-function arises from the discontinuity at $t=t'$. A system with a time independent hamiltonian $[h_{sr}]$ gives a propagator according to Equation (3.15) which is a function of the time interval $t-t'$.

It is often convenient to study the temporal behaviour of the propagator $G_{sr}(t-t')$ from Equation (3.15) in terms of its Fourier integral, which is defined by the expression

$$G_{sr}(t-t')=(2\pi)^{-1}\int_{-\infty}^{\infty} dE\, G_{sr}(E)\exp[-iE(t-t')]. \tag{3.19}$$

The inverse relation

$$G_{sr}(E)=\int_{-\infty}^{\infty} d(t-t')G_{sr}(t-t')\exp[iE(t-t')], \tag{3.20}$$

leads to improper integrals of the type

$$\int_0^{\infty} dt \exp[it(E-\epsilon_k)]$$

which may be dealt with by the introduction of a convergence factor $e^{-t\eta}$ in the integrand and by taking the limit when η tends to zero on the positive real axis after integration.

We obtain from Equations (3.15) and (3.20) that

$$G_{sr}(E)=\lim_{\eta\to+0}\sum_k x_{sk}\left[\frac{f_k}{E-\epsilon_k-i\eta}+\frac{1-f_k}{E-\epsilon_k+i\eta}\right]x^\dagger_{kr}. \tag{3.21}$$

This expression allows us to define $G_{sr}(E)$ for arbitrary complex E as

$$G_{sr}(E)=\sum_k x_{sk}[E-\epsilon_k]^{-1}x_{kr}^{\dagger} \tag{3.22}$$

which is an element of the resolvent to the hamiltonian matrix $[h_{rs}]$:

$$EG_{sr}(E)=\delta_{sr}+\sum_q h_{sq}G_{qr}(E). \tag{3.23}$$

It may be used in the integral (3.19) instead of the form (3.21) if the integration of the energy variable is performed as a contour integral in the complex E-plane. The contour is chosen in such a way as to bypass the singularities at $E=\epsilon_k$ on the real axis. An acceptable contour is pictured in Fig. 1 for the case that $f_k \geq f_l$ when $\epsilon_k \leq \epsilon_l$.

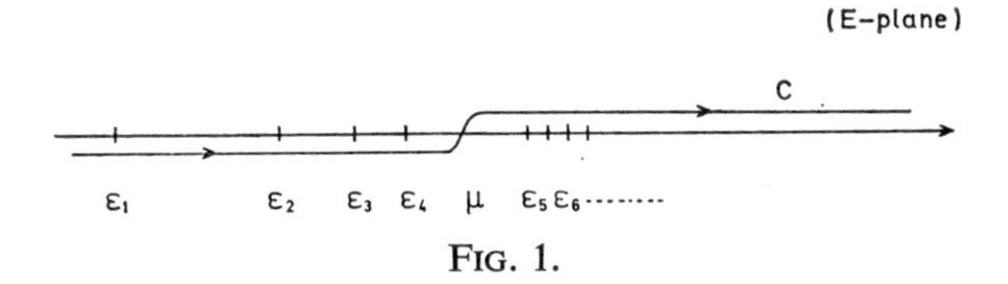

FIG. 1.

The conditions on the occupation numbers imply that this corresponds to the ground state of the N-electron system and that there is a parameter μ such that

$$f_k=\begin{cases}1, & \epsilon_k<\mu,\\ 0, & \epsilon_k>\mu.\end{cases} \tag{3.24}$$

A continuous spectrum can be handled similarly when considered as the limit of a discrete one.

The Fourier transform $G_{sr}(E)$ of the propagator is an analytic function in the complex E-plane except on the real axis where it can have simple poles and cuts. These correspond to energy eigenvalues of the single particle hamiltonian. We investigate the discontinuity of the function $G_{sr}(E)$ at the real axis and consider

$$\begin{aligned}A_{sr}(\epsilon)&=\lim_{\eta\to+0}(i/2\pi)\,[G_{sr}(\epsilon+i\eta)-G_{sr}(\epsilon-i\eta)],\\ &=\sum_k x_{sk}x_{kr}^{\dagger}\lim_{\eta\to+0}(\eta/\pi)/[(\epsilon-\epsilon_k)^2+\eta^2]\\ &=\sum_k x_{sk}x_{kr}^{\dagger}\delta(\epsilon-\epsilon_k).\end{aligned} \tag{3.25}$$

The Dirac δ-function appears here as the limiting function

$$\delta(x)=\lim_{\eta\to 0}(\eta/\pi)/[x^2+\eta^2], \tag{3.26}$$

which satisfies all criteria for that "function", particularly

$$\int_a^b \delta(x-c)f(x)dx = \begin{cases} f(c), & \text{if } a<c<b \\ 0, & \text{otherwise.} \end{cases} \tag{3.27}$$

The function $A_{sr}(\epsilon)$, defined in Equation (3.25), is called the spectral density function. The knowledge of $A_{sr}(\epsilon)$ suffices to yield $G_{sr}(E)$ in the complex E-plane from the integral

$$G_{sr}(E) = \int_{-\infty}^{\infty} d\epsilon \, A_{sr}(\epsilon)/[E-\epsilon]. \tag{3.28}$$

The physical properties that can be calculated from the Green's function can thus be computed once the spectral density function is known.

The considerations in this chapter have centered around the interpretation of and formal relations for propagators. Before we discuss a particular example we give two equations that relate stationary state properties of an N electron system to the propagator.

It follows from Equation (3.15) that

$$\begin{aligned} N = \sum f_k &= \lim_{t \to t'-0} -i\sum_s G_{ss}(t,t') \\ &= (2\pi i)^{-1} \int_C dE \, \sum_s G_{ss}(E) \\ &= \sum_s \int_{-\infty}^{\mu} d\epsilon A_{ss}(\epsilon). \end{aligned} \tag{3.29}$$

The contour C is the one in Fig. 1 plus an infinite semicircle in the upper half of the complex E-plane or any other closed contour obtainable from it by deformations which do not bring the integration path through singularities. A particular choice is the contour used by Coulson and shown in Fig. 2.

The total energy can be calculated similarly as

$$\begin{aligned} E_0 = \sum_k \epsilon_k f_k &= \lim_{t \to t'-0} \sum_s \partial G_{ss}(t,t')/\partial t \\ &= (2\pi i)^{-1} \int_C E dE \, \sum_s G_{ss}(E) \\ &= \sum_s \int_{-\infty}^{\mu} \epsilon d\epsilon A_{ss}(\epsilon), \end{aligned} \tag{3.30}$$

with the same choice of contour as in Equation (3.29). These relations are examples of the initial conditions that specify the Green's function in conjunction with the differential equations. They appear as sum rules for the spectral density function. Other physical properties will be derived from the electron propagator in the following chapters.

The Hückel Model

E. Hückel introduced a quantum mechanical model for the description of the electronic structure of planar unsaturated molecules that has since been

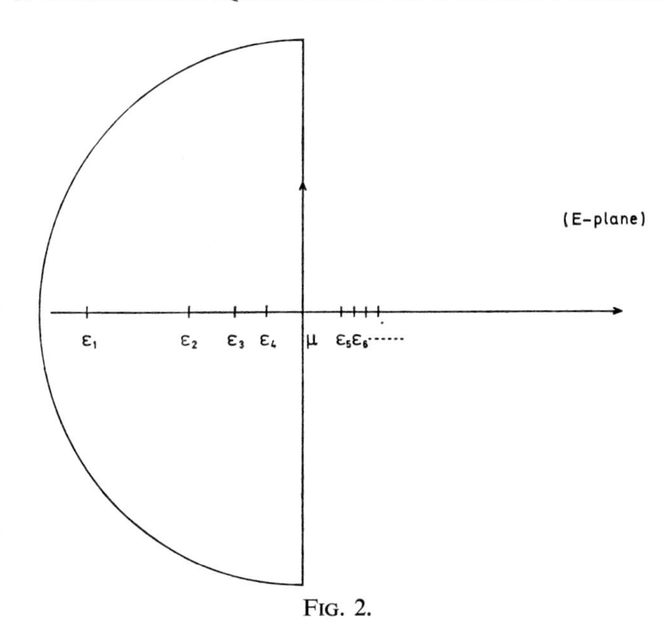

FIG. 2.

widely used. Coulson and Longuet–Higgins gave the theory a very elegant formulation which connects readily to a Green's function treatment. The hamiltonian is independent of spin. Consequently we can limit the discussion to the situation where the labels r, s, and so on refer to orbitals rather than to spin orbitals. Each orbital is associated with a particular atomic site in the molecule, and with one orbital per site there is a direct correspondence between indexes and sites.

We work directly with the Fourier transform of the electron propagator or the resolvent, and a matrix notation is used such that

$$\mathbf{G}(E)=[G_{sr}(E), s, r=1, 2, \ldots M].$$

Equation (3.23) then reads

$$(E\mathbf{1}-\mathbf{h})\mathbf{G}(E)=\mathbf{1}. \tag{3.31}$$

The secular determinant,

$$D(E)= \| E\mathbf{1}-\mathbf{h} \|, \tag{3.32}$$

is a function of the various matrix elements h_{sr} and from the theory of matrices one obtains that

$$G_{sr}(E)=(-)^{r+s+1}\, \partial \log D(E)/\partial h_{rs}. \tag{3.33}$$

This formula is valid when all elements h_{rs} are considered to be independent, that is

$$\partial h_{pq}/\partial h_{rs}=\delta_{rp}\delta_{sq}. \tag{3.34}$$

It holds also that

$$Tr\,\mathbf{G}(E)=\sum_s G_{ss}(E)=\partial \log D(E)/\partial E. \tag{3.35}$$

For the simple case of doubly filled orbitals we can now express the total number of electrons and the total electronic energy as

$$N=(\pi i)^{-1}\int_C dE\, \partial \log D(E)/\partial E, \tag{3.36}$$

and

$$E_0=(\pi i)^{-1}\int_C E dE\, \partial \log D(E)/\partial E, \tag{3.37}$$

respectively.

The particular feature of the Hückel model are the choices of matrix elements where

$$h_{rr}=\alpha_r \tag{3.38}$$

and

$$h_{rs}=h_{sr}=\beta_{sr}, \qquad s\neq r, \tag{3.39}$$

and all elements are real. The β-parameters are zero unless the sites r and s are neighbors. The formal electron charge on site s is defined as

$$q_s=\partial E_0/\partial \alpha_s, \tag{3.40}$$

and may equivalently be expressed as

$$q_s=(\pi i)^{-1}\int_C dE\, G_{ss}(E), \tag{3.41}$$

which readily follows from Equations (3.37) and (3.33).

Similarly we obtain the mobile bond order,

$$p_{sr}=\tfrac{1}{2}\, \partial E_0/\partial \beta_{sr}, \tag{3.42}$$

as the integral

$$\begin{aligned} p_{sr}&=(2\pi i)^{-1}\int_C dE\, [G_{sr}(E)+G_{rs}(E)] \\ &=(\pi i)^{-1}\int_C dE\, G_{sr}(E). \end{aligned} \tag{3.43}$$

Higher derivatives of the energy have also been introduced and are called mutual atom–atom, atom–bond, and bond–bond polarizabilities. They are all obtainable from the general second derivative

$$\partial^2 E_0/\partial h_{rs}\partial h_{s'r'}=(\pi i)^{-1}\int_C dE\, G_{ss'}(E)G_{r'r}(E), \tag{3.44}$$

where the form of the derivative of $\mathbf{G}(E)$ with respect to any parameter comes from the differentiation of Equation (3.31).

Problem 1

Demonstrate Coulson's energy formula,

$$E_0=Tr\,\mathbf{h}+\pi^{-1}\int_{-\infty}^{\infty} dy\, [(\mu+iy)D'(\mu+iy)/D(\mu+iy)-M],$$

through the use of the contour integral expression (3.37) and the asymptotic form of $D(E)$ for large E-values.

Problem 2

The Hückel problem for planar unsaturated hydrocarbons leads to the parameter choices,

$$h_{rr} = \alpha_r = \alpha = \mu,$$

$$h_{rs} = h_{sr} = \beta_{sr}, \text{ for bonded atoms } s \text{ and } r,$$

$$h_{rs} = 0, \text{ otherwise,}$$

and the propagator equation is

$$(E-\mu)G_{sr}(E) = \delta_{sr} + \sum \beta_{sq}G_{qr}(E).$$

Show that iteration of this equation gives

$$[(E-\mu)^2 - \sum |\beta_{rq}|^2]\, G_{rr}(E) = (E-\mu) + \sum_q \sum_{p \neq r} \beta_{rq}\beta_{qp}G_{pr}(E).$$

Discard the last term on the right hand side and calculate the total energy and bond orders for molecules where all β's are equal. Show that the average energy per atom for a graphite layer equals $\mu + \beta(3)^{\frac{1}{2}}$ in this approximation, which is less than 10% in error with respect to the accurate value $\mu + 1{\cdot}576\beta$.

Notes and Bibliography

An excellent exposition of the propagator concept and its exploitation in quantum mechanics is given by R. P. Feynman and A. R. Hibbs (1965) in "Quantum Mechanics and Path Integrals", McGraw-Hill Book Co., New York. Several useful formulae in the theory of matrices, determinants and resolvents are developed by R. Courant and D. Hilbert (1953) in "Methods of Mathematical Physics", Vol. 1, Interscience Publishers, New York. An elegant form of Hückel theory was formulated by C. A. Coulson and H. C. Longuet-Higgins (1947) in a series of papers in *Proc. Roy. Soc.* **A191**, 39; **A192**, 16; **A193**, 447; **A193**, 456. Coulson published his integration method in 1940 in *Proc. Camb. Phil. Soc.* **40**, 201.

CHAPTER 4

Fermion Operators

The material in the preceding chapter will now be formalized in order to find a way to consider interacting electrons. We develop an algebra for the study of many electrons starting from the assumption that the coefficients $[a_s(t)]$ in the expansion $\psi(\xi t)=\sum_s u_s(\xi)a_s(t)$ are operators rather than numbers. They are not hermitian and we consider their adjoints $[a_s^\dagger(t)]$ as well. The algebra arises from the postulate that operators referring to equal times satisfy anticommutation relations:

$$\begin{aligned}[][a_s, a_r]_+ &= a_s a_r + a_r a_s = 0,\\ [a_s^\dagger, a_r^\dagger]_+ &= a_s^\dagger a_r^\dagger + a_r^\dagger a_s^\dagger = 0,\\ [a_s, a_r^\dagger]_+ &= a_s a_r^\dagger + a_r^\dagger a_s = \delta_{rs}.\end{aligned} \tag{4.1}$$

The operators are generators of a Grassman algebra, but we will not consider this aspect. Equation (4.1) has the consequence that

$$(a_r)^2=(a_r^\dagger)^2=0.$$

It also follows from the operator nature of the a's that $\psi(\xi t)$ itself is an operator which has the anticommutation relations

$$\begin{aligned}[][\psi(\xi t), \psi(\xi' t)]_+ &= 0,\\ [\psi^\dagger(\xi t), \psi^\dagger(\xi' t)]_+ &= 0,\\ [\psi(\xi t), \psi^\dagger(\xi' t)]_+ &= \delta(\xi-\xi').\end{aligned} \tag{4.2}$$

These algebraic rules will be shown to be sufficient to determine a basis in the Hilbert space on which the operators act. This corresponds closely to the algebraic methods for the analysis of angular momentum and harmonic oscillator eigenvalues and eigenstates. Quantum mechanics based on operator algebra is most reasonably treated in the Heisenberg picture, where state vectors are time independent and operators develop in time according to the Heisenberg equation of motion. Thus we have that the operator $a_s(t)$ should satisfy

$$i\,da_s/dt=[a_s, H] \qquad (\hbar=1) \tag{4.3}$$

and this can only be equivalent with Equation (3.2) if the hamiltonian operator is

$$H = \sum a_s^\dagger h_{sr} a_r \tag{4.4}$$

or equivalently

$$\begin{aligned} H &= \int d\xi \psi^\dagger(\xi) h \psi(\xi) \\ &= \int d\xi \psi^\dagger(\xi) \left[(-i\nabla - e\bar{A}/c)^2/2m - e\bar{\sigma}.\bar{B}/2mc + V(\xi)\right] \psi(\xi), \end{aligned} \tag{4.5}$$

where $\bar{B} = \text{curl } \bar{A}$, $\bar{A}$ is the vector potential, and $\bar{\sigma}$ are the Pauli spin operators. This form of an operator is adapted for situations with electromagnetic fields present.

Other operators of interest are, for instance, the electronic charge and current density operators

$$q(\bar{r}) = e\sum_{\text{spin}} \psi^\dagger(\xi)\, \psi(\xi), \tag{4.6}$$

$$\begin{aligned} \bar{j}(\bar{r}) = (e/2m) \sum_{\text{spin}} &[(i\nabla\psi^\dagger(\xi))\, \psi(\xi) - i\psi^\dagger(\xi)\nabla\psi(\xi) \\ &- 2\psi^\dagger(\xi)\, (e\bar{A}(\bar{r}, t)/c)\psi(\xi) + \text{curl } \psi^\dagger(\xi)\bar{\sigma}\psi(\xi)] \end{aligned} \tag{4.7}$$

where the last sum in Equation (4.7) derives from the spin density. The total electronic charge in the system will be given by the eigenvalues of the operator

$$Q = \int d\bar{r}\, q(\bar{r}) = e \int d\xi \psi^\dagger(\xi)\, \psi(\xi) = eN_{\text{op}}. \tag{4.8}$$

The electron number operator N_{op} is thus given by

$$N_{\text{op}} = \sum a_s^\dagger a_s = \sum n_s \tag{4.9}$$

and we shall examine it to gain some further insight in our algebra.

It follows from Equation (4.1) that

$$[n_s, n_r] = 0, \tag{4.10}$$

and that

$$n_s^2 = a_s^\dagger a_s a_s^\dagger a_s = a_s^\dagger(1 - a_s^\dagger a_s) a_s = n_s - a_s^\dagger a_s^\dagger a_s a_s = n_s, \tag{4.11}$$

which means that the eigenvalues of n_s are 0 and 1. Thus we conclude that N_{op} has the natural numbers and zero as eigenvalues. We postulate that the state with eigenvalue zero is a nondegenerate state. It is called the vacuum state:

$$N_{\text{op}} \mid \text{vac}\rangle = 0. \tag{4.12}$$

We consider the states

$$\mid r\rangle = a_r^\dagger \mid \text{vac}\rangle$$

and find that

$$n_s \mid r\rangle = \delta_{sr} \mid r\rangle \tag{4.13}$$

and

$$N_{\text{op}} \mid r\rangle = \mid r\rangle \tag{4.14}$$

i.e. $\mid r\rangle$ is a one-electron state.

Similarly we generate states

$$|rs\rangle = a_r^\dagger a_s^\dagger|\text{ vac}\rangle,\ |rr\rangle = 0, \tag{4.15}$$

and find that

$$n_p|rs\rangle = (\delta_{pr} + \delta_{ps})|rs\rangle, \tag{4.16}$$

as well as

$$N_{\text{op}}|rs\rangle = 2|rs\rangle, \tag{4.17}$$

i.e. $|rs\rangle$ is a two-electron state.

The states

$$|rs\ldots\rangle = a_r^\dagger a_s^\dagger \ldots |\text{ vac}\rangle \tag{4.18}$$

form a basis for our description and we note that there will occur linear dependencies if we employ all possible products. The simplest case is

$$|rs\rangle = -|sr\rangle$$

from Equation (4.1).

Our state vectors are generated by the electron creation operators $a_r^\dagger$ and we can deduce that the adjoint operators a_r serve to step down the ladder we have built:

$$a_r|rs\rangle = |s\rangle,\ a_s|rs\rangle = -|r\rangle.$$

The a_r's are called annihilation operators for electrons. The eigenstates of the occupation number operators n_r are not stationary states unless the hamiltonian (4.4) is given in diagonal form:

$$H = \sum_r \epsilon_r a_r^\dagger a_r = \sum_r \epsilon_r n_r. \tag{4.19}$$

It is always possible to transform a general expression of the form (4.4) into an equivalent expression (4.19) since it follows from Equation (3.8) that

$$H = \sum_{sr} a_s^\dagger h_{sr} a_r = \sum_k (\sum_s a_s^\dagger x_{sk}) \epsilon_k (\sum_r x_{kr}^\dagger a_r) \tag{4.20}$$

and with the notation

$$\tilde{a}_k = \sum_r x_{kr}^\dagger a_r, \tag{4.21}$$

we find that the transformed operators $\tilde{a}_k$ also satisfy the anticommutation relations (4.1), while the hamiltonian takes the form

$$H = \sum_k \tilde{a}_k^\dagger \epsilon_k \tilde{a}_k = \sum \epsilon_k \tilde{n}_k. \tag{4.22}$$

Eigenstates of the occupation number operators $\tilde{n}_k$ are stationary states when the hamiltonian (4.22) applies. The normal state with N electrons occupying the N lowest energy spin orbitals is given by the expression

$$|0\rangle = \prod_k [\tilde{a}_k^\dagger]^{f_k}|\text{vac}\rangle = \prod_k [\sum_r a_r^\dagger x_{rk}]^{f_k}|\text{vac}\rangle, \tag{4.23}$$

where f_k is given by Equation (3.13). This state gives the relations

$$(1-f_k)\tilde{a}_k|0\rangle=0, \tag{4.24}$$

$$f_k\tilde{a}_k^\dagger|0\rangle=0. \tag{4.25}$$

These equations express compactly that electrons can neither be removed from unoccupied spin orbitals nor put into occupied ones.

The Electron Propagator

We will now prove that the Green's function $G_{sr}(t, t')$ which was discussed in the previous chapter is given, in the second quantization language, as

$$G_{sr}(t, t')=\begin{cases}-i\langle 0\,|a_s(t)a_r^\dagger(t')|\,0\rangle, & t>t',\\ i\langle 0\,|a_r^\dagger(t')a_s(t)|\,0\rangle, & t<t'.\end{cases} \tag{4.26}$$

The operators develop in time according to Equation (4.3) which can be solved in the manner of Equation (3.2) to give

$$a_s(t)=\sum_k x_{sk}\exp(-i\epsilon_k t)\tilde{a}_k(0). \tag{4.27}$$

It follows from Equation (4.24) that

$$a_s(t)\,|0\rangle=\sum_k x_{sk}\exp(-i\epsilon_k t)f_k\tilde{a}_k(0)|\,0\rangle, \tag{4.28}$$

and from Equation (4.25) that

$$\langle 0\,|a_s(t)=\sum_k x_{sk}\exp(-i\epsilon_k t)\langle 0\,|\tilde{a}_k(0)\,(1-f_k). \tag{4.29}$$

Equations (4.28) and (4.29) and the similar results for the operator $a_r^\dagger(t')$ lead to

$$G_{sr}(t, t')=\sum_k x_{sk}G_k(t, t')x_{kr}^\dagger, \tag{4.30}$$

with the notation

$$G_k(t, t')=\begin{cases}-i\langle 0\,|\tilde{a}_k(t)\tilde{a}_k^\dagger(t')\,|0\rangle=-i(1-f_k)\exp[-i\epsilon_k(t-t')], & t>t',\\ i\langle 0\,|\tilde{a}_k^\dagger(t')\tilde{a}_k(t)|\,0\rangle=if_k\exp[-i\epsilon_k(t-t')], & t<t'.\end{cases} \tag{4.31}$$

The interpretation of the Green's function in the present notation will be preserved in the theory of interacting electrons where the ground state is more complicated than the simple expression (4.23).

Slater Determinants

We have, up to this point, discussed the quantum mechanical many electron problem in terms of a basis $[u_s(\xi)]$. It was noted in Equation (4.2) that $\psi(\xi)$

and $\psi^\dagger(\xi)$ are operators and we might consider, in conjunction with the states of Equation (4.18), states of the form

$$|\xi_1, \xi_2, \ldots, \xi_N\rangle = \psi^\dagger(\xi_1)\psi^\dagger(\xi_2) \ldots \psi^\dagger(\xi_N)|\,\text{vac}\rangle. \tag{4.32}$$

Such a state is a combination of the states of Equation (4.18) and we are interested in the scalar products

$$\begin{aligned}\langle \xi_1\, \xi_2 \ldots \xi_N\, | r_1\, r_2 \ldots r_N \rangle \\ &= \langle \text{vac}\, |\psi(\xi_N)\psi(\xi_{N-1}) \ldots \psi(\xi_1) a^\dagger_{r_1} a^\dagger_{r_2} \ldots a^\dagger_{r_N}|\,\text{vac}\rangle \\ &= \det |u_{r_1}(\xi_1) u_{r_2}(\xi_2) \ldots u_{r_N}(\xi_N)|. \end{aligned} \tag{4.33}$$

The Slater determinants of Equation (4.33) appear in the formalism of second quantization as expansion coefficients of a spin orbital configuration $|r_1\, r_2 \ldots r_N\rangle$ in terms of spatial configurations $|\xi_1 \ldots \xi_N\rangle$. The derivation of Equation (4.33) rests on the anticommutation relation

$$[\psi(\xi), a^\dagger_r]_+ = u_r(\xi). \tag{4.34}$$

Electron Interaction

The hamiltonian (4.4) or (4.5) applies only to non-interacting electrons and it is necessary to consider its generalization to the case of an interacting system of electrons. We approach this problem by adding to the energy operator (4.5) the Coulomb interaction energy of the charge density $q(\bar{r})$ of Equation (4.6) with itself:

$$\begin{aligned} H'_{\text{int}} &= \tfrac{1}{2} \int d\bar{r}\, d\bar{r}'\, q(\bar{r}) q(\bar{r}')\, |\bar{r} - \bar{r}'|^{-1} \\ &= \tfrac{1}{2} \textstyle\sum_{rsr's'} (rs\, | r's') a^\dagger_r a_s a^\dagger_{r'} a_{s'}. \end{aligned} \tag{4.35}$$

The matrix elements are given as

$$(rs|r's') = e^2 \int d\xi\, d\xi' u^*_r(\xi) u_s(\xi)\, |\bar{r} - \bar{r}'|^{-1} u^*_{r'}(\xi') u_{s'}(\xi'). \tag{4.36}$$

The H'_{int} is not correct since it gives a non-vanishing expectation value for a one-electron state where no interaction energy should occur:

$$\langle r\, |H'_{\text{nt}}|\, r\rangle = \textstyle\sum_s (rs|sr). \tag{4.37}$$

This is an inadmissible part of H'_{int}. We rewrite Equation (4.35) as follows:

$$H'_{\text{int}} = \tfrac{1}{2} \textstyle\sum_{rsr's'} (rs|r's') a^\dagger_r a^\dagger_{r'} a_{s'} a_s + \tfrac{1}{2} \textstyle\sum_{rss'} (rs|ss') a^\dagger_r a_{s'}$$

and disregard the last term to define

$$\begin{aligned} H_{\text{int}} &= \tfrac{1}{2} \textstyle\sum_{rsr's'} (rs|r's') a^\dagger_r a^\dagger_{r'} a_{s'} a_s \\ &= (e^2/2) \int d\xi\, d\xi' \psi^\dagger(\xi) \psi^\dagger(\xi')\, |\bar{r} - \bar{r}'|^{-1} \psi(\xi') \psi(\xi), \end{aligned} \tag{4.38}$$

as the appropriate term to be added to the hamiltonian for non-interacting electrons. The derivation of H_{int} from H'_{int} is the formation of the normal product of operators and amounts to permuting creation operators to the right and annihilation operators to the left while accounting for sign changes according to the anticommutation relations and discarding terms resulting from the Kronecker deltas of Equation (4.1).

The Heisenberg equation of motion for the operator a_s is no longer linear when interactions are included

$$i\,da_s/dt = \sum_r h_{sr} a_r + \sum_{r's'r} (sr\,|s'r')\, a_{s'}^{\dagger} a_{r'} a_r. \tag{4.39}$$

The correlation effects of the electronic motion are exhibited in the last terms of Equation (4.39) and it will be the main problem in the following to represent these terms approximately in order to allow for the electronic interactions.

There are two identities which we want to employ in the calculations. They are

$$\sum_s a_s^{\dagger}[a_s, H] = \sum_{sr} a_s^{\dagger} h_{sr} a_r = H, \tag{4.40}$$

and

$$\sum_s a_s^{\dagger}[a_s, H_{\text{int}}] = \sum_{rsr's'} (sr\,|s'r')\, a_s^{\dagger} a_{s'}^{\dagger} a_{r'} a_r = 2H_{\text{int}}. \tag{4.41}$$

The total hamiltonian H_{tot},

$$H_{\text{tot}} = H + H_{\text{int}}, \tag{4.42}$$

then satisfies

$$\sum_s a_s^{\dagger}[a_s, H_{\text{tot}}] = H + 2H_{\text{int}}. \tag{4.43}$$

There is a formal similarity of the commutator $[a_s, H_{\text{tot}}]$ with the functional derivative of H_{tot} with respect to $a_s^{\dagger}$, which is useful to observe.

Operators of Angular Momentum

The generators of the three-dimensional rotation group are in normal unrelativistic quantum mechanics given as

$$\bar{j} = \bar{l} + \bar{s}, \tag{4.44}$$

where

$$\bar{l} = \bar{r} \times \bar{p} = -i\bar{r} \times \nabla \tag{4.45}$$

is the operator for orbital angular momentum and

$$\bar{s} = \tfrac{1}{2}\bar{\sigma} \tag{4.46}$$

is the operator of the spin angular momentum, expressed in the Pauli spin matrices. The operator for the total angular momentum for a many-electron system is then constructed in analogy with the previous sections as

$$\bar{J} = \int \psi^{\dagger}(\xi)\bar{j}\psi(\xi)\,d\xi = \sum_{rs} (\bar{j})_{rs} a_r^{\dagger} a_s. \tag{4.47}$$

This operator can obviously be written as a sum of orbital and spin angular momentum operators.

Many molecular hamiltonians commute with the total spin operator which leads us to consider some transformation properties of field operators under rotations in spin space. It is then natural to employ a basis of the form

$$u_{s\pm\frac{1}{2}}(\xi)=\chi_s(\bar{r})\begin{cases}\alpha(\zeta)\\ \beta(\zeta)\end{cases} \tag{4.48}$$

where α and β are the eigenvectors the z-component of $\bar{s}$, and the orbitals $[\chi_s(\bar{r})]$ form an orthonormal set. Let unit vectors in the three cartesian coordinate directions be $\bar{e}_x$, $\bar{e}_y$, $\bar{e}_z$. We obtain for the operator of total spin

$$\bar{S}=\int \psi^\dagger(\xi)\bar{s}\psi(\xi)d\xi=\textstyle\sum_r\bar{S}_r, \tag{4.49}$$

where

$$\bar{S}_r=\bar{e}_xS_{rx}+\bar{e}_yS_{ry}+\bar{e}_zS_{rz}$$

and

$$\begin{aligned}S_{rx}&=\tfrac{1}{2}(a^\dagger_{r+\frac{1}{2}}a_{r-\frac{1}{2}}+a^\dagger_{r-\frac{1}{2}}a_{r+\frac{1}{2}}),\\ S_{ry}&=-\tfrac{1}{2}i(a^\dagger_{r+\frac{1}{2}}a_{r-\frac{1}{2}}-a^\dagger_{r-\frac{1}{2}}a_{r+\frac{1}{2}}),\\ S_{rz}&=\tfrac{1}{2}(n_{r+\frac{1}{2}}-n_{r-\frac{1}{2}}).\end{aligned} \tag{4.50}$$

We consider now a rotation in spin space about an axis the direction and magnitude of which are given by a vector $\bar{\Theta}$. The transformed field operators are then

$$\begin{aligned}\tilde{a}^\dagger_{r\nu}&=\exp(i\bar{\Theta}.\bar{S})a^\dagger_{r\nu}\exp(-i\bar{\Theta}.\bar{S})\\ &=\exp(i\bar{\Theta}.\bar{S}_r)a^\dagger_{r\nu}\exp(-i\bar{\Theta}.\bar{S}_r),\end{aligned} \tag{4.51}$$

where the last line follows since the operators $\bar{\Theta}.\bar{S}_r$ commute. We obtain the commutation relations ($\Theta=|\bar{\Theta}|$)

$$\begin{aligned}[\bar{\Theta}.\bar{S}_r, a^\dagger_{r\nu}]&=\nu\Theta_z a^\dagger_{r\nu}+(\tfrac{1}{2}\Theta_x+i\nu\Theta_y)a^\dagger_{r-\nu},\\ [\bar{\Theta}.\bar{S}_r,[\bar{\Theta}.\bar{S}_r, a^\dagger_{r\nu}]]&=(\tfrac{1}{2}\Theta)^2a^\dagger_{r\nu},\end{aligned} \tag{4.52}$$

which lead us to the result that

$$\begin{aligned}\tilde{a}^\dagger_{r\nu}&=a^\dagger_{r\nu}+i[\bar{\Theta}.\bar{S}_r, a^\dagger_{r\nu}]-\tfrac{1}{2}[\bar{\Theta}.\bar{S}_r,[\bar{\Theta}.\bar{S}_r, a^\dagger_{r\nu}]]+\dots\\ &=a^\dagger_{r\nu}\cos\tfrac{1}{2}\Theta+i[\bar{\Theta}.\bar{S}_r, a^\dagger_{r\nu}]\,(\tfrac{1}{2}\Theta)^{-1}\sin\tfrac{1}{2}\Theta.\end{aligned} \tag{4.53}$$

This is consistent with the two-dimensional irreducible representations of the three-dimensional rotation group. We encounter in later chapters further applications of these properties.

Problem 1

Demonstrate that the continuity equation $dq(\bar{r})/dt+\operatorname{div}\bar{j}(\bar{r})=0$, holds for the charge and current density operators in second quantization.

Problem 2

Calculate a parameter λ such that the two electron state $|rs\rangle$ is obtained as a unitary transformation of the vacuum state:

$$|rs\rangle = \exp\,(i\lambda a_r^\dagger a_s^\dagger + i\lambda^* a_s a_r)|\text{ vac}\rangle \qquad (\lambda = -\tfrac{1}{2}i\pi).$$

Notes and Bibliography

Second quantization was introduced for the handling of many-particle systems already in the early days of quantum mechanics by Born, Jordan, Dirac, Fock and others. Those who are intrigued by the mathematical aspects of the field operator algebra are referred to F. A. Berezin, "The Method of Second Quantization" (Academic Press, New York, 1966). The transformation properties of field operators have been discussed by F. A. Kaempffer in "Concepts in Quantum Mechanics" (Academic Press, New York, 1965).

CHAPTER 5

Double-time Green's Functions

Formal Definitions

The definition of the electron propagator in Equation (4.26) is a special case of the general concept of a double-time Green's function. More generally one attempts to calculate the effect on an observation, represented by an operator A, by another observation represented by B. When the observations are performed at different times the general Green's function is denoted as

$$\langle\langle A(t); B(t') \rangle\rangle.$$

Particular examples of such Green's functions are response functions or influence functions, among them dynamic polarizabilities, conductivities and susceptibilities.

The Green's functions are defined here with reference to the density operator ρ for an ensemble which may be grand canonical, canonical, microcanonical, or pure state. We will only be concerned with situations at absolute zero temperature, but we will in the formalism leave the possibility open also for the case of non-zero temperature. Thus we introduce a density operator, ρ, which is normalized such that

$$Tr\rho = 1. \tag{5.1}$$

Operators will be called fermion-like if they are odd with respect to the transformation $\psi(\xi) \rightarrow -\psi(\xi)$, and if they are even they are called boson-like. The Green's function is defined as

$$\langle\langle A(t); B(t') \rangle\rangle = -i\theta(t-t')\langle A(t)B(t')\rangle \pm i\theta(t'-t)\langle B(t')A(t)\rangle \tag{5.2}$$

where the upper sign in the second term occurs only if both A and B are fermion-like operators. The Heaviside function $\theta(t)$ is the integral of the Dirac δ-function

$$\theta(t) = \int_{-\infty}^{t} \delta(\tau)d\tau. \tag{5.3}$$

Average values like those occurring on the right hand side of Equation (5.2) are defined as

$$\langle A(t)B(t') \rangle = Tr\rho A(t)B(t'). \tag{5.4}$$

The operator $A(t)$ satisfies the Heisenberg equation of motion

$$i\frac{d}{dt}A(t) = [A(t), H] \tag{5.5}$$

and the Green's function will then satisfy the differential equation

$$i\frac{d}{dt}\langle\langle A(t); B(t')\rangle\rangle = \delta(t-t')\langle A(t)B(t') \pm B(t')A(t)\rangle + \langle\langle [A(t), H]; B(t')\rangle\rangle. \tag{5.6}$$

The density operator ρ and the hamiltonian H will generally be constant in time and the Green's function will be a function of the time interval $(t-t')$ only. It is then often advantageous to attempt a solution of Equation (5.6) by means of a Fourier transform. Thus we define

$$\langle\langle A; B\rangle\rangle_E = \int_{-\infty}^{\infty} d(t-t')\langle\langle A(t); B(t')\rangle\rangle e^{+iE(t-t')}, \tag{5.7}$$

and obtain from Equations (5.6) and (5.7)

$$E\langle\langle A; B\rangle\rangle_E = \langle AB \pm BA\rangle + \langle\langle [A, H]; B\rangle\rangle_E. \tag{5.8}$$

The differential equation (5.6) and the algebraic equation (5.8) need to be supplemented with initial conditions, the actual nature of which will be discussed in connection with specific problems.

We proceed to study the spectral representations of the Green's functions under the assumption that the density operator ρ is diagonal in the basis of the eigenstates of the hamiltonian H:

$$\langle n|\rho|m\rangle = \rho_n\delta_{nm}. \tag{5.9}$$

Matrix elements of the operators A and B are obtained from Equation (5.5) as

$$\langle n|A(t)|m\rangle = \langle n|A|m\rangle e^{-it(E_m-E_n)}. \tag{5.10}$$

The definition (5.2) then leads to the form

$$\begin{aligned}\langle\langle A(t); B(t')\rangle\rangle &= -i\theta(t-t')\sum_{nm}\rho_n\langle n|A(t)|m\rangle\langle m|B(t')|n\rangle\\ &\quad \pm i\theta(t'-t)\sum_{nm}\rho_m\langle m|B(t')|n\rangle\langle n|A(t)|m\rangle\\ &= \sum_{nm}\langle n|A|m\rangle\langle m|B|n\rangle e^{-i(t-t')(E_m-E_n)}\\ &\quad \times[-i\rho_n\theta(t-t') \pm i\rho_m\theta(t'-t)].\end{aligned} \tag{5.11}$$

This function has the Fourier transform [cf. Equation (3.21)]

$$\langle\langle A; B\rangle\rangle_E = \sum_{nm}\langle n|A|m\rangle\langle m|B|n\rangle \times\left[\frac{\rho_n}{E-E_m+E_n+i\eta} \pm \frac{\rho_m}{E-E_m+E_n-i\eta}\right]. \tag{5.12}$$

A pure state density operator is characterized by the relation

$$\rho_n = \delta_{n0}, \tag{5.13}$$

and then the Green's function simplifies to

$$\langle\langle A; B\rangle\rangle_E = \sum_m \left[\frac{\langle 0\,|A|\,m\rangle\,\langle m\,|B|\,0\rangle}{E-E_m+E_0+i\eta} \pm \frac{\langle 0\,|B|\,m\rangle\,\langle m\,|A|\,0\rangle}{E+E_m-E_0-i\eta}\right]. \tag{5.14}$$

The spectral representation (5.12) exhibits the properties of the Green's function in a particularly lucid way. Singularities occur at values of the energy parameter E that equal energy differences for stationary states. The residues at the poles give information on transition amplitudes between stationary states.

The Electron Propagator

Following the definition (4.26) we obtain the electron propagator for the most general case as

$$G_{sr}(t-t') = \langle\langle a_s(t); a_r^\dagger(t')\rangle\rangle \tag{5.15}$$

with the Fourier transform

$$G_{sr}(E) = \langle\langle a_s; a_r^\dagger\rangle\rangle_E. \tag{5.16}$$

The spectral resolution of $G_{sr}(E)$ as given by the expression (5.12) demonstrates that contributions to the sum arise from states such that

$$\langle m\,|a_r^\dagger|\,n\rangle \neq 0. \tag{5.17}$$

This implies that for a hamiltonian which commutes with the number operator the stationary state $|\,m\rangle$ will contain one electron more than the state $|\,n\rangle$. Thus we conclude that the corresponding energy difference E_m-E_n relates to the energy change when an electron is added. A pure state electron propagator will give information on the energy required to remove an electron or the amount of released energy for the process of adding an electron. This generalization is in complete accord with the notions imbedded in the description of the electron propagator in Chapter 3 where interactions are neglected. We find that there is a correspondence of matrix elements $\langle m\,|a_r^\dagger|\,n\rangle$ and eigenvector elements $x_{kr}^\dagger$ and of orbital energies ϵ_k and differences E_m-E_n. A comparison of Equations (4.30) and (5.11) with use of Equation (4.31) yields that

$$\langle m\,|\tilde{a}_k^\dagger|\,n\rangle = 1, \qquad E_m-E_n=\epsilon_k, \tag{5.18}$$

such that for a pure state either

$$\langle m\,|\tilde{a}_k^\dagger|\,0\rangle = 1, \qquad E_m=E_0+\epsilon_k, \tag{5.19}$$

or

$$\langle 0\,|\tilde{a}_k^\dagger|\,n\rangle = 1, \qquad E_n=E_0-\epsilon_k. \tag{5.20}$$

The propagator also contains information on ground state properties as already indicated in Chapter 3. We see that according to the definitions we have that

$$\begin{aligned}\langle\psi^\dagger(\xi)\psi(\xi)\rangle &= -i \lim_{t'\to t+0} \langle\langle\psi(\xi t);\, \psi^\dagger(\xi t')\rangle\rangle \\ &= (2\pi i)^{-1}\int_C dE\langle\langle\psi(\xi);\, \psi^\dagger(\xi)\rangle\rangle_E,\end{aligned} \tag{5.21}$$

where the contour consists of the real axis and an infinite semicircle in the upper half plane as in Fig. 1 (Chapter 3). Thus the electron density is related to the initial conditions of the Green's function and in particular we can relate the one particle reduced density matrix to the propagator as in Equation (3.43):

$$p_{sr} = \langle a_r^\dagger a_s\rangle = (2\pi i)^{-1}\int_C dE\langle\langle a_s;\, a_r^\dagger\rangle\rangle_E, \tag{5.22}$$

where s and r now label spin orbitals and summation over spin is not included. The calculation of the total energy of the system can also be related to the electron propagator as in Chapter 3, but the presence of the electron interaction requires some modifications which will be exemplified in the particular applications in later chapters.

The Polarizability Tensor

Another important Green's function is obtained when the operators A and B are identified with the dipole moment operator

$$\bar{R} = \int \bar{r}q(\bar{r})d\bar{r}. \tag{5.23}$$

The Green's function will now occur as a tensor with nine components $\langle\langle\bar{R};\, \bar{R}\rangle\rangle_E$, which will be identified with the negative of the frequency dependent polarizability tensor in Chapter 12. This Green's function can be calculated if all Green's functions of the form $\langle\langle a_r^\dagger a_s;\, a_{s'}^\dagger a_{r'}\rangle\rangle_E$ are known. It follows from the spectral representation Equation (5.12) that singularities will occur for energies $E = E_m - E_n$ such that the number of electrons is the same in states $|\, m\rangle$ and $|\, n\rangle$. Transitions between such states are important in the process of absorption and scattering of electromagnetic radiation. A discussion of these phenomena is given in Chapter 12.

Problem

Derive an expression for the pure state propagator

$$\langle\langle a_r^\dagger a_s;\, a_m^\dagger a_n\rangle\rangle_E$$

when the ground state is the independent particle state

$$|0\rangle = \prod_k\, [\tilde{a}_k^\dagger]^{f_k}|\ \text{vac}\rangle$$

of Equation (4.23), i.e. the N lowest spin orbitals with energies ϵ_1, ϵ_2, . . . ϵ_k, . . . ϵ_N are occupied.

Notes and Bibliography

D. N. Zubarev has written an interesting account of the theory and application of general double-time Green's functions, published in 1960 in *Soviet Physics Uspekhi* **3** 320 (English translation).

CHAPTER 6

Simple Applications

It is instructive to consider a few simple examples where the abstract theory is elucidated through a detailed treatment. Non-interacting electrons in a constant potential and in a centrally symmetric potential are studied, Green's functions with appropriate boundary conditions are constructed, and from them some physical properties are derived.

Electrons in a Box

The electron propagator is given as the integral

$$\langle\langle \psi(\xi t);\, \psi^\dagger(\xi' t')\rangle\rangle = (2\pi)^{-1}\int dE\, \mathrm{e}^{-iE(t-t')} G(\xi, \xi'; E), \tag{6.1}$$

and with the hamiltonian

$$H = \int d\xi\, \psi^\dagger(\xi)\, [-\nabla^2/2m]\, \psi(\xi) \tag{6.2}$$

we obtain the equation of motion

$$[E + \nabla^2/2m]\, G(\xi, \xi'; E) = \delta(\bar{r} - \bar{r}')\delta_{\zeta\zeta'}, \tag{6.3}$$

where $\xi = (\bar{r}, \zeta)$ is the compound space–spin coordinate. The boundary conditions on the propagator will now be important for the solution of Equation (6.3). A simple case occurs when the only requirement is that the Green's function should vanish when the distance $|\bar{r} - \bar{r}'| = \rho$ is infinitely large. We choose then the wave number k as

$$k = (2mE)^{\frac{1}{2}}, \qquad \mathrm{Im}\, k > 0. \tag{6.4}$$

The resulting propagator is

$$G(\xi, \xi'; E) = -\delta_{\zeta\zeta'}(m/2\pi\rho)\mathrm{e}^{ik\rho}. \tag{6.5}$$

This Green's function has a branch cut along the positive real E-axis showing the continuous energy spectrum of a free particle.

We suppose for a moment that Equation (6.3) is given for one spatial coordinate only and that the Green's function should vanish outside the interval $a \leq x \leq b$. Following the general procedure in Chapter 2 we then obtain the result

$$G(\xi, \xi'; E) = \delta_{\zeta\zeta'} G(x, x'; E),$$
$$G(x, x'; E) = [2m/k \sin k(b-a)] \sin k(x_{<} - a) \sin k(x_{>} - b). \quad (6.6)$$

The singularities of G are simple poles at

$$E = \frac{1}{2m}\left(\frac{2\pi n}{b-a}\right)^2, \qquad n = 1, 2, 3, \ldots \quad (6.7)$$

in accordance with the well-known spectrum for a particle in a one dimensional box of length $b-a$.

The solution for the cases of two- and three-dimensional rectangular boxes can now be simply obtained through convolution integrals. Thus we have in two dimensions

$$G(xy, x'y'; E) = (2\pi i)^{-1} \int_C d\epsilon \, G(x, x'; \epsilon) \, G(y, y'; E-\epsilon) \quad (6.8)$$

where the contour C starts at $+\infty$ and encircles the positive real axis counterclockwise and ends at $+\infty$. The contour should have all poles of $G(x, x'; \epsilon)$ inside and all poles of $G(y, y'; E-\epsilon)$ outside, both Green's functions given by expressions like those in Equation (6.6). Correspondingly, in three dimensions, the result is

$$G(xyz, x'y'z'; E) = (2\pi i)^{-1} \int d\epsilon \, G(xy, x'y'; \epsilon) \, G(z, z'; E-\epsilon), \quad (6.9)$$

where Equations (6.6) and (6.8) are used.

We notice that while the eigenfunctions for a separable hamiltonian can be expressed as products of eigenfunctions for the separate parts, the Green's functions will be convolution integrals of Green's functions for the various terms in the hamiltonian.

Electrons in a Spherical Potential

The electron propagator for a spherically symmetric spin independent potential is given as in Equation (6.1) with

$$G(\xi, \xi'; E) = \delta_{\zeta\zeta'} G(\bar{r}, \bar{r}'; E). \quad (6.10)$$

Only three variables are significant for $G(\bar{r}, \bar{r}'; E)$ and we choose them as

$$r = |\bar{r}|,$$
$$r' = |\bar{r}'|,$$

and γ where

$$\bar{r} . \bar{r}' = rr' \cos \gamma.$$

The hamiltonian is separable in the variables r and γ and we obtain the total Green's function as a convolution:

$$G(\bar{r}, \bar{r}'; E) = (2\pi i)^{-1} \int_C d\lambda\, D(\cos\gamma; \lambda) G_\lambda(r, r'; E)/rr'. \tag{6.11}$$

The angular Green's function satisfies the equation

$$[\lambda + \partial^2/\partial\gamma^2 + \cot\gamma\,\partial/\partial\gamma]\, D(\cos\gamma; \lambda) = \delta(1-\cos\gamma), \tag{6.12}$$

and has the spectral form

$$D(\cos\gamma; \lambda) = \sum_{l=0}^{\infty} (2l+1)\, P_l(\cos\gamma)/4\pi\, [\lambda - l(l+1)], \tag{6.13}$$

in terms of the Legendre polynomials.

The radial Green's function is obtained from the equation

$$[2m(E-V(r)) - \lambda/r^2 + \partial^2/\partial r^2]\, G_\lambda(r, r'; E) = 2m\delta(r-r'), \tag{6.14}$$

and the boundary conditions that it vanishes at the origin. The contour C is again such that it surrounds the positive real λ-axis in the counter-clockwise fashion such that no poles of G_λ fall inside the contour.

We can obtain a closed expression for $D(\cos\gamma; \lambda)$ from the theory of Chapter 2 and the knowledge of the Legendre functions $P_\nu(\cos\gamma)$ which are solutions to

$$[\nu(\nu+1) + \partial^2/\partial\gamma^2 + \cot\gamma\,\partial/\partial\gamma]\, P_\nu(\cos\gamma) = 0. \tag{6.15}$$

Contrary to the situation for the Legendre polynomials, the Legendre functions for general complex degree ν are not regular for $\gamma = \pi$. It holds that $P_\nu(1) = 1$ and we obtain two linearly independent solutions to Equation (6.15), $P_\nu(\cos\gamma)$ and $P_\nu(-\cos\gamma) = P_\nu(\cos(\pi-\gamma))$, which are finite at the end points of the interval, $\gamma = 0$ and $\gamma = \pi$, respectively.

The Green's function $D(\cos\gamma; \nu(\nu+1))$ satisfies Equation (6.15) when $0 < \gamma \leq \pi$ and is here proportional to $P_\nu(-\cos\gamma)$ since it follows from the spectral form (6.13) that it has a finite value at $\gamma = \pi$. We obtain the constant of proportionality by noticing that according to Equation (6.13)

$$D(\cos\pi; \nu(\nu+1)) = \pi/\sin\nu\pi, \tag{6.16}$$

and thus

$$D(\cos\gamma; \nu(\nu+1)) = \pi P_\nu(-\cos\gamma)/\sin\nu\pi. \tag{6.17}$$

It follows that the Green's function is singular for integer values of ν and the residues are found to agree with those from the spectral form (6.13). The relation between ν and λ is

$$\nu = -\tfrac{1}{2} \pm (\lambda + \tfrac{1}{4})^{\frac{1}{2}}, \tag{6.18}$$

and the contour integral (6.11) can be changed into one involving the variable ν where the contour then surrounds the real ν-axis in the counter-clockwise direction.

There are but few potentials $V(r)$ which admit closed form solutions of Equation (6.14) in terms of well-known functions. Among them is the Coulomb case where $V(r) = -e^2Z/r$ and the Green's function has been obtained by several authors in various representations.

We derive an approximate solution to Equation (6.14) for a general potential $V(r)$ by means of the Wentzel–Kramers–Brillouin method and demonstrate how the solution can be employed to calculate quantities of interest. Kramer's modification of the differential equation for the spherical case is used in that λ is replaced by $(\nu+\frac{1}{2})^2$. Thus we define the local wave number as

$$p_\nu(r) = [2m(E - V(r)) - (\nu+\tfrac{1}{2})^2 r^{-2}]^{\frac{1}{2}}, \qquad \operatorname{Im} p_\nu(r) > 0. \tag{6.19}$$

The two linearly independent solutions of the homogeneous equation corresponding to Equation (6.14) are given in the WKB-approximation as

$$\exp\left[\pm i \int^r p_\nu(s)\, ds - \tfrac{1}{2} \log p_\nu(r)\right].$$

Their wronskian equals $2i$ and the Green's function, which should vanish when r or r' equals zero or infinity, is then

$$G_\nu(r, r'; E) = (m/i) \exp\left[i \int_{r_<}^{r_>} p_\nu(s) ds - \tfrac{1}{2} \log p_\nu(r) - \tfrac{1}{2} \log p_\nu(r')\right], \tag{6.20}$$

where ν has replaced λ as the subscript. The notations $r_>$ and $r_<$ stand for the greater and lesser of r and r' respectively and the integral in the exponent is thus performed over a positive interval which together with the choice of branch for the local wave number guarantees the desired boundary conditions.

The Green's function obtained in Equation (6.20) has, for real ν, a branch cut along the real E-axis thus exhibiting a completely continuous spectrum and we can determine its spectral density function as defined in Chapter 3.

We write the complex energy variable as $E = \epsilon + i\eta$ and define

$$k_\nu(r) = |2m(\epsilon - V(r)) - (\nu+\tfrac{1}{2})^2 r^{-2}|^{\frac{1}{2}}. \tag{6.21}$$

It follows then that when r is a point in a so-called forbidden region, that is

$$\epsilon < V(r) + (\nu+\tfrac{1}{2})^2/(2mr^2), \tag{6.22}$$

then

$$\lim_{\eta \to 0} p_\nu(r) = ik_\nu(r). \tag{6.23}$$

When r is a point in an allowed region, that is

$$\epsilon > V(r) + (\nu+\tfrac{1}{2})^2/(2mr^2), \tag{6.24}$$

then

$$\lim_{\eta\to\pm 0} p_\nu(r) = \pm k_\nu(r). \tag{6.25}$$

The spectral density function is, according to Equation (3.25),

$$A_\nu(r, r'; \epsilon) = \lim_{\eta\to+0} (i/2\pi)\,[G_\nu(r, r'; \epsilon+i\eta) - G_\nu(r, r'; \epsilon-i\eta)]. \tag{6.26}$$

Several distinct cases can be recognized depending upon whether r and r' are in forbidden or allowed regions and the number of turning points, zeros of $k_\nu(r)$, in the interval (r, r'). We assume for simplicity that r is less than r'.

Case (i). r and r' both in forbidden region, no turning points in between.

$$A_\nu(r, r'; \epsilon) = 0. \tag{6.27}$$

Case (ii). r and r' both in forbidden regions, two turning points, $a<b$, in between.

$$A_\nu(r, r'; \epsilon) = (m/\pi)[k_\nu(r)k_\nu(r')]^{-\frac{1}{2}} \times \exp\left[-\int_r^a k_\nu(s)ds - \int_b^{r'} k_\nu(s)ds\right] \sin\left[\int_a^b k_\nu(s)ds\right]. \tag{6.28}$$

Case (iii). r in forbidden and r' in allowed region, one turning point, a, in between.

$$A_\nu(r, r'; \epsilon) = (m/\pi)\,[k_\nu(r)k_\nu(r')]^{-\frac{1}{2}} \exp\left[-\int_r^a k_\nu(s)ds\right] \times \sin\left[\int_a^{r'} k_\nu(s)ds + \pi/4\right]. \tag{6.29}$$

Case (iv). r and r' in allowed region, no turning points in between.

$$A_\nu(r, r'; \epsilon) = (m/\pi)\,[k_\nu(r)k_\nu(r')]^{-\frac{1}{2}} \cos\left[\int_r^{r'} k_\nu(s)ds\right]. \tag{6.30}$$

Case (v). r and r' in allowed regions, two turning points $a<b$, in between.

$$A_\nu(r, r'; \epsilon) = (m/\pi)\,[k_\nu(r)k_\nu(r')]^{-\frac{1}{2}} \exp\left[-\int_a^b k_\nu(s)ds\right] \times \cos\left[\int_r^a k_\nu(s)ds + \int_b^{r'} k_\nu(s)ds\right]. \tag{6.31}$$

These formulae can be collected in the form

$$A_\nu(r, r'; \epsilon) = (m/\pi)\,[k_\nu(r)k_\nu(r')]^{-\frac{1}{2}} \exp\,[-\alpha] \sin\beta \tag{6.32}$$

where the exponent α is the sum of integrals of k_ν over the forbidden regions between r and r' and β is the sum of integrals of k_ν over the allowed regions plus $\pi/4$ for each end point in an allowed region.

The spectral density function $A_\nu(r, r'; \epsilon)$ is clearly related to the wave functions in the WKB-approximation. Particularly we notice that in Case (iii) it is obtained as the product of the two connecting solutions across a turning point. The Green's function does not, at the present level of approximation, give the Bohr–Sommerfeld quantization condition, but we observe that in Case (ii) we obtain the spectral density function as the product of the wave functions in the end points of the interval and an amplitude factor which has its extremum when the quantization condition,

$$\int_a^b k_\nu(s)ds = (n+\tfrac{1}{2})\pi, \tag{6.33}$$

holds.

A very simple expression is obtained for the electron density when the WKB-approximation is employed for the Green's function. The radial density distribution for real angular momentum ν is given as

$$n_\nu(r) = (2\pi i)^{-1} \int_C dE\, G_\nu(r, r; E) = (2\pi i)^{-1} \int_C dE\, [m/ip_\nu(r)], \tag{6.34}$$

which is analogous to Equation (3.41). A change of integration variable from E to p_ν allows us to conclude that

$$\begin{aligned} n_\nu(r) &= -(2\pi)^{-1} \int dp_\nu = k_\nu(r)/\pi, \quad && \epsilon = \mu > V(r) + (\nu+\tfrac{1}{2})^2 r^{-2} \\ &= 0, && \text{otherwise.} \end{aligned} \tag{6.35}$$

This formula is the one used by Fermi to obtain the number of electrons in a given angular momentum state in the statistical theory of atoms. We observe that there is an upper bound to the possible angular momentum in that the radial density distribution vanishes when

$$|\nu+\tfrac{1}{2}| > rk_{-\frac{1}{2}}(r), \tag{6.36}$$

corresponding to r times the maximum linear momentum.

The total density is similarly obtained from

$$\begin{aligned} n(\bar{r}) &= (2\pi i)^{-1} \int_C dE\, \Sigma_\xi G(\xi, \xi; E) = (\pi i)^{-1} \int_C dE\, G(\bar{r}, \bar{r}; E) \\ &= \Sigma_l (l+\tfrac{1}{2}) n_l(r)/\pi r^2. \end{aligned} \tag{6.37}$$

We may approximate the last summation by an integration over l so that

$$n(\bar{r}) \simeq -(\pi/3) \int d[n_l(r)]^3 = (\pi/3)\, [n_{-\frac{1}{2}}(r)]^3, \tag{6.38}$$

where the integration limits on l have been taken as $-\frac{1}{2}$ and $rk_{-\frac{1}{2}}(r)$. This result allows us to relate the parameter μ, which separates the occupied and unoccupied levels, to the density,

$$\mu = V(r) + (2m)^{-1}\, [3\pi^2\, n(\bar{r})]^{2/3}, \tag{6.39}$$

which is the result for an electron gas and reduces to one of the conditions for the statistical model of the atom when μ equals zero. These results indicate that μ has the role of the chemical potential.

We can calculate the total number of electrons from the density given by Equation (6.38),

$$N=4\pi\int n(\bar{r})r^2dr, \tag{6.40}$$

and obtain for the case of a Coulomb potential $V(r)=-e^2Z/r$, that

$$N=(2/3)\,[-e^4Z^2m/2\mu]^{3/2}. \tag{6.41}$$

Conversely we have that the parameter μ is given as

$$\mu=-\tfrac{1}{2}e^4Z^2m\,[2/3N]^{2/3}, \tag{6.42}$$

which implies that the principal quantum number of the highest occupied level equals the integer part of $[3N/2]^{1/3}$.

Problem 1

Construct the spectral representation of the propagator for an electron in a one-dimensional box by considering the residues at the poles of the closed expression $G(x, x'; E)$ of Equation (6.6) and derive the form of the normalized wave functions for the stationary states.

Problem 2

Consider the Green's function for the one-dimensional harmonic oscillator,

$$[E-\tfrac{1}{2}x^2+\tfrac{1}{2}d^2/dx^2]\,G(x, x'; E)=\delta(x-x'),$$

and derive a closed expression from the theory of parabolic cylinder functions. A comprehensive list of formulae is given by M. Abramowitz and I. A. Stegun, "Handbook of Mathematical Functions", National Bureau of Standards, Applied Mathematics Series, 55 (1964) Chapter 19.

Problem 3

Calculate the energy density for the WKB-approximation to the electron propagator as

$$W_\nu(r)=(2\pi i)^{-1}\int_C E\,dE\,G_\nu(r, r; E).$$

Notes and Bibliography

The Green's function for a free particle is obtained in standard texts on quantum mechanics, for instance L. I. Schiff, "Quantum Mechanics", 2nd ed. (McGraw-Hill Book Co., New York, 1955), in connection with scattering problems. The Coulomb Green's function was obtained in closed form by L. Hostler and R. H. Pratt,

Phys. Rev. Letters **10**, 469 (1963). Several other representations are available, notably the one given by J. Schwinger in *J. Math. Phys.* **5**, 1606 (1964).

The connection between the WKB-approximation for atomic wave functions and the statistical model of the atom has been given by J. S. Plaskett in *Proc. Roy. Soc.* **A66**, 178 (1953), and A. A. Broyles in *Am. J. Phys.* **29**, 81 (1961). R. Gáspár has given the relation between the number of electrons and the principal quantum number of the highest occupied energy level in a Coulomb field in *Acta Physica Academiae Scientiarum Hungaricae* **27**, 441 (1969). The calculation by Fermi of the number of s-, p-, and d-electrons is discussed by Condon and Shortley in "The Theory of Atomic Spectra" (Cambridge, 1963).

CHAPTER 7

Aspects of the Hartree–Fock Approximation

Physical arguments were used by Hartree to establish the self-consistent field model for atoms. Slater and Fock demonstrated the relation to the variational method and brought in the exchange potential as a consequence of the fermion character of electrons. Later developments have given various alternative approaches to the same basic equations and we attempt to review here those aspects that are related to the Green's function approach to many-electron problems.

According to the equation of motion for the electron propagator we have that

$$G_{sr}(E) = \langle\langle a_s; a_r^\dagger \rangle\rangle_E = E^{-1} [\langle [a_s, a_r^\dagger]_+ \rangle + \langle\langle [a_s, H]; a_r^\dagger \rangle\rangle_E], \tag{7.1}$$

where the hamiltonian is

$$H = \sum_{sr} h_{sr} a_s^\dagger a_r + \tfrac{1}{2} \sum_{s'sr'r} (s'r'|sr) a_s^\dagger a_{s'}^\dagger a_{r'} a_r. \tag{7.2}$$

We also assume that the basis is orthonormal. The interaction terms in the hamiltonian generate terms on the right hand side of Equation (7.1) that cannot simply be expressed in terms of G_{sr}'s. It is now an important problem to decide whether a reasonable approximation can be made such that

$$\langle\langle [a_s, H]; a_r^\dagger \rangle\rangle_E \cong \sum_q f_{sq} G_{qr}(E). \tag{7.3}$$

When this can be accomplished the problem is reduced to solving the equation

$$\sum_q (E\delta_{sq} - f_{sq}) G_{qr}(E) = \delta_{sr}. \tag{7.4}$$

The notations from Chapter 3 will now be used so that the eigenvalues and eigenvectors of the matrix $[f_{sq}]$ are ϵ_k and $[x_{sk}]$ respectively, with the relation

$$\sum_q f_{sq} x_{qk} = \epsilon_k x_{sk}. \tag{7.5}$$

We require the matrix $[f_{sq}]$ to be hermitian.

The solution of Equation (7.4) for a complex E-value is

$$G_{sr}(E) = \sum_k x_{sk} (E - \epsilon_k)^{-1} x_{kr}, \tag{7.6}$$

as in Equation (3.22) which should be compared to the general spectral representation of Equation (5.12). Thus we can identify the residues as

$$\langle n | a_s | m \rangle \langle m | a_r^\dagger | n \rangle (\rho_n + \rho_m) = x_{sk} x_{kr}^\dagger, \tag{7.7}$$

and the poles as

$$E_m - E_n = \epsilon_k. \tag{7.8}$$

We examine first the case when the density operator is that of a grand canonical ensemble so that

$$\rho_m = \exp\,[\beta(F - E_m + \mu N_m)], \tag{7.9}$$

where β^{-1} is the absolute temperature times Boltzmann's constant, F is the free energy, μ is the chemical potential, and N_m is the number of electrons in state m. It follows that

$$\rho_n = \rho_m \exp\,[\beta(\epsilon_k - \mu)], \tag{7.10}$$

and we conclude that

$$\langle a_r^\dagger a_s \rangle = \Sigma_k x_{sk} x_{kr}^\dagger / (1 + \exp\,[\beta(\epsilon_k - \mu)]), \tag{7.11}$$

where we recognize the Fermi–Dirac distribution for non-interacting electrons with an energy spectrum $[\epsilon_k]$. Particularly we obtain

$$\langle \tilde{n}_k \rangle = \langle \tilde{a}_k^\dagger \tilde{a}_k \rangle = 1/(1 + \exp\,[\beta(\epsilon_k - \mu)]), \tag{7.12}$$

for the operators defined by Equation (4.21). The Green's function for real E-values can now be specified in terms of these initial values as

$$G_{sr}(E) = \Sigma_k x_{sk} \left[\frac{\langle \tilde{n}_k \rangle}{E - \epsilon_k - i\eta} + \frac{1 - \langle \tilde{n}_k \rangle}{E - \epsilon_k + i\eta} \right] x_{kr}^\dagger, \tag{7.13}$$

similar to the form (3.21) with the limiting process implicitly understood.

The expression (7.13) for the electron propagator is also the most general one consistent with the equation of motion (7.4). This is important to realize since it makes it possible to specify a set of average values $\langle \tilde{n}_k \rangle$ corresponding to some ensemble which is not necessarily defined in terms of thermodynamical parameters. We will see that for example the choice

$$\rho = \Pi_k\,[1 - \langle \tilde{n}_k \rangle - \tilde{n}_k + 2\,\langle \tilde{n}_k \rangle\, \tilde{n}_k], \tag{7.14}$$

is such that

$$Tr\,\rho = 1, \tag{7.15}$$

and

$$Tr\,\tilde{n}_k \rho = \langle \tilde{n}_k \rangle, \tag{7.16}$$

which is a consistent definition of the ensemble.

Derivation of the Fock Operator from a Moment Expansion

The equation of motion (7.1) for the propagator can be iterated to read

$$G_{sr}(E) = E^{-1}\,\langle [a_s, a_r^\dagger]_+ \rangle + E^{-2}\,\langle [\,[a_s, H], a_r^\dagger]_+ \rangle \ldots . \tag{7.17}$$

Let us identify this with the corresponding form derived from the spectral density function $A_{sr}(\epsilon)$:

$$G_{sr}(E) = \int d\epsilon A_{sr}(\epsilon)\,[E^{-1} + E^{-2}\epsilon + \ldots], \tag{7.18}$$

which is derived from Equation (3.28). Thus we can view Equation (7.17) as a moment expansion for the distribution function $A_{sr}(\epsilon)$ and by comparison of terms in Equations (7.4), (7.17), and (7.18) we find

$$\delta_{sr} = \langle [a_s, a_r^{\dagger}]_+ \rangle = \int d\epsilon\; A_{sr}(\epsilon), \tag{7.19}$$

$$f_{sr} = \langle [[a_s, H], a_r^{\dagger}]_+ \rangle = \int \epsilon d\epsilon\; A_{sr}(\epsilon), \tag{7.20}$$

and so on.

We can now calculate a more detailed expression for the matrix elements f_{sr} from the formulae given in Chapter 4:

$$f_{sr} = h_{sr} + \textstyle\sum_{s'r'}[(sr\,|s'r') - (sr'|\,s'r)]\,\langle a_{s'}^{\dagger}\, a_{r'} \rangle, \tag{7.21}$$

where the one-particle reduced density matrix occurs. This expression has the form given by Fock for the effective one-particle hamiltonian and the matrix $[f_{sr}]$ will henceforth be called the Fock matrix. It opens up the possibility for a self-consistent determination of the propagator since the reduced density matrix can be computed according to Equations (5.22) and (7.13), from the Green's function, which is given in terms of the Fock matrix in Equation (7.4).

Such a procedure as described above gives a slightly more general approach to the Hartree–Fock equations than the ordinary wave function derivation and allows the definition of a self-consistent potential from an ensemble with specified occupation numbers. Particularly it should be noticed that a case where all numbers $\langle \tilde{n}_k \rangle$ of Equations (7.13) and (7.14) equal zero or unity corresponds to an ensemble containing only one state with a configuration space representative in the form of a single Slater determinant. We have reason to explore more general choices of occupation numbers in connection with the occurrence of partially occupied degenerate energy levels as for instance in open shell atoms, which are treated in Chapter 8.

An approximation of the propagator in terms of the truncated moment expansion correct through the first moment will yield a discontinuity of the correct magnitude of its first time-derivative at equal times. This indicates that the Hartree–Fock method will be adequate for description of short time processes where relaxation phenomena are unimportant. This is also to be seen in the definition of the Fock matrix with its universal potential term.

The elements of the Fock matrix appears as first moments of the spectral density function $A_{sr}(\epsilon)$ for the electron propagator. Their determination through a self-consistency procedure will also have consequences for the higher moments. In particular we obtain the geometric progression

$$G_{sr}(E)=E^{-1}\delta_{sr}+E^{-2}f_{sr}+E^{-3}\sum_q f_{sq}f_{qr}+\dots. \tag{7.22}$$

The second moment equals, according to Equation (7.1),

$$\int \epsilon^2 d\epsilon A_{sr}(\epsilon)=\langle[[[a_s, H], H], a_r^\dagger]_+\rangle, \tag{7.23}$$

and we can compare this with the coefficient of E^{-3} in Equation (7.22).

Let us define the operator b_s as the difference

$$b_s=[a_s, H]-\sum_q f_{sq}a_q. \tag{7.24}$$

Then we have that

$$\begin{aligned}\langle[[[a_s, H], H], a_r^\dagger]_+\rangle &=\sum_q f_{sq}f_{qr}+\langle[[b_s, H], a_r^\dagger]_+\rangle\\ &=\sum_q f_{sq}f_{qr}+\langle[b_s, b_r^\dagger]_+\rangle,\end{aligned} \tag{7.25}$$

and we see that the matrix elements f_{sq} are such that the positive quantity $\langle[b_s, b_s^\dagger]_+\rangle$ has a minimum. The Hartree–Fock approximation has, in this sense, an optimum property also with regard to the second moment.

The Variational Method

We wish to prove in this section that the Hartree–Fock approximation has stationary properties with regard to the calculation of the total energy of a many-particle system. The expectation value of the total hamiltonian requires the calculation of average values $\langle a_r^\dagger a_s\rangle$ and $\langle a_r^\dagger a_{r'}^\dagger a_{s'}a_s\rangle$ from the ensemble (7.14). The density operator,

$$\rho=\Pi_k(c_k+d_k\tilde{n}_k), \tag{7.26}$$

$$c_k=1-\langle\tilde{n}_k\rangle, \tag{7.27}$$

$$d_k=2\langle\tilde{n}_k\rangle-1, \tag{7.28}$$

is a direct product operator and each factor is an operator in a two-dimensional space, which in particular could be spanned by the eigenstates of $\tilde{n}_k$. Then we have the following matrix representatives of the operators:

$$\tilde{n}_k=\begin{pmatrix}0&0\\0&1\end{pmatrix},\ \tilde{a}_k=\begin{pmatrix}0&1\\0&0\end{pmatrix},\ \tilde{a}_k^\dagger=\begin{pmatrix}0&0\\1&0\end{pmatrix}. \tag{7.29}$$

The trace of a direct product operator is the product of the traces of the individual factors and we find for instance that

$$Tr\rho=\Pi_k[c_k\, Tr1+d_k\, Tr\tilde{n}_k]=\Pi_k(2c_k+d_k)=1, \tag{7.30}$$

as was stated in Equation (7.15). Similarly we obtain

$$\langle \tilde{n}_k \rangle = Tr\, \tilde{n}_k \rho = c_k + d_k, \tag{7.31}$$

in agreement with Equations (7.16), (7.27), and (7.28). The latter result can be generalized to read

$$\langle \tilde{a}_k^\dagger \tilde{a}_l \rangle = Tr\, \tilde{a}_k^\dagger \tilde{a}_l \rho = \delta_{kl} \langle \tilde{n}_k \rangle. \tag{7.32}$$

Expectation values of products of four fermion operators are

$$\langle \tilde{a}_k^\dagger \tilde{a}_l^\dagger \tilde{a}_{l'} \tilde{a}_{k'} \rangle = [\delta_{kk'} \delta_{ll'} - \delta_{kl'} \delta_{lk'}] \langle \tilde{n}_k \rangle \langle \tilde{n}_l \rangle, \tag{7.33}$$

since the trace of a single $\tilde{a}_k$-operator is zero and there are two ways of pairing off the annihilation and creation operators in Equation (7.33).

From the results of Equations (7.32) and (7.33) we find the required expectation values as

$$\langle a_r^\dagger a_s \rangle = \textstyle\sum_k x_{sk} \langle \tilde{n}_k \rangle x_{kr}^\dagger = p_{sr}, \tag{7.34}$$

and

$$\begin{aligned} \langle a_r^\dagger a_{r'}^\dagger a_{s'} a_s \rangle &= \textstyle\sum x_{sk} x_{s'l} \langle \tilde{n}_k \rangle \langle \tilde{n}_l \rangle [x_{lr'}^\dagger x_{kr}^\dagger - x_{kr'}^\dagger x_{lr}^\dagger] \\ &= p_{sr} p_{s'r'} - p_{sr'} p_{s'r}, \end{aligned} \tag{7.35}$$

where the inverse relation of Equation (4.21) has been used.

The average value of the hamiltonian with the use of the ensemble (7.26) equals

$$\begin{aligned} \langle H \rangle &= \textstyle\sum h_{rs} p_{sr} + \frac{1}{2} \sum (rs\,|r's') [p_{sr} p_{s'r'} - p_{sr'} p_{s'r}] \\ &= \textstyle\sum h_{rs} p_{sr} + \frac{1}{2} \sum [(rs\,|r's') - (rs'|\,r's)]\, p_{sr} p_{s'r'} \\ &= \textstyle\frac{1}{2} \sum [h_{rs} + f_{rs}]\, p_{sr}. \end{aligned} \tag{7.36}$$

The energy expression depends only upon the matrix elements p_{sr} which are the elements of the Fock–Dirac reduced density matrix. It follows from Equation (7.36) upon variation that

$$\delta \langle H \rangle = \textstyle\sum f_{rs} \delta p_{sr} = Tr\, \mathbf{f} \delta \mathbf{p}, \tag{7.37}$$

in matrix notation.

A general unitary transformation of the basis which conserves the factorization property (7.35) and thus the independent particle character of the ensemble is considered in the form

$$U = \exp [i\Lambda] = \exp [i \textstyle\sum_{sr} \lambda_{sr} a_s^\dagger a_r], \tag{7.38}$$

$$\Lambda = \Lambda^\dagger,$$

so that the hermiticity of Λ ensures the unitarity of U. The new density operator will be taken as

$$\rho' = U^\dagger \rho U = \rho + i[\rho, \Lambda] - \tfrac{1}{2}[[\rho, \Lambda], \Lambda] + \ldots, \tag{7.39}$$

and we consider infinitesimal transformations such that

$$\begin{aligned}\delta p_{sr} &= i Tr[\rho, \Lambda]\, a_r^\dagger a_s = i Tr \rho[\Lambda, a_r^\dagger a_s] \\ &= \Sigma_q [p_{sq}\lambda_{qr} - \lambda_{sq} p_{qr}] = i[\mathbf{p}, \boldsymbol{\lambda}]_{sr}.\end{aligned} \tag{7.40}$$

The result (7.40) is inserted into Equation (7.37) to obtain

$$\delta\langle H\rangle = i Tr \mathbf{f}\, [\mathbf{p}, \boldsymbol{\lambda}] = i Tr\, [\mathbf{f}, \mathbf{p}]\, \boldsymbol{\lambda} = 0, \tag{7.41}$$

since **f** and **p** are commuting matrices.

Stability Properties of Hartree–Fock Calculations

It is proven in the preceding section that the first variation of the energy in the Hartree–Fock approximation vanishes. The related question of characterization of the stationary point requires a treatment of the second variation which will be carried out in this section. At the same time we can explore the stability of the solution towards external perturbations represented by one-electron operators in the hamiltonian. In the latter case we are led to consider variations also in the h_{sr}'s, so that

$$\begin{aligned}\delta\langle H\rangle &= \Sigma\, [f_{rs}\delta p_{sr} + \delta h_{rs} p_{sr}] \\ &= \Sigma\, \delta h_{rs} p_{sr} = \langle \delta H\rangle,\end{aligned} \tag{7.42}$$

in accordance with the Hellman–Feynman theorem. The second variation is

$$\delta^2\langle H\rangle = \langle \delta^2 H\rangle + \tfrac{1}{2}\Sigma(\delta h_{rs} + \delta f_{rs})\, \delta p_{sr} + \Sigma\, f_{rs}\delta^2 p_{sr}, \tag{7.43}$$

with

$$\delta f_{sr} = \delta h_{sr} + \Sigma\, [(sr\,|s'r') - (sr'\,|s'r)]\, \delta p_{r's'}, \tag{7.44}$$

from Equation (7.21).

A derivation of the second variation of the reduced density matrix from Equation (7.39) gives

$$\delta^2 \mathbf{p} = -\tfrac{1}{2}[[\mathbf{p}, \boldsymbol{\lambda}], \boldsymbol{\lambda}], \tag{7.45}$$

analogously to the result (7.40). We are now prepared to express the second variation of the energy entirely in terms in the λ's. It simplifies matters to employ the basis which diagonalizes **f** and **p**. Thus we obtain

$$\begin{aligned}\delta^2\langle H\rangle = \langle \delta^2 H\rangle &+ i\,\Sigma\, \delta h_{kl}\, [\langle \tilde{n}_l\rangle - \langle \tilde{n}_k\rangle]\, \lambda_{lk} \\ &- \tfrac{1}{2}\Sigma\, [(lk\,|l'k') - (lk'|\,l'k)]\, [\langle \tilde{n}_k\rangle - \langle \tilde{n}_l\rangle]\, [\langle \tilde{n}_{k'}\rangle - \langle \tilde{n}_{l'}\rangle]\, \lambda_{kl}\lambda_{k'l'} \\ &- \Sigma\, \epsilon_k [\langle \tilde{n}_k\rangle - \langle \tilde{n}_l\rangle]\, |\lambda_{kl}|^2.\end{aligned} \tag{7.46}$$

The further analysis is facilitated by the use of more compact notations. Each pair of indices (kl) is given the label ν and the corresponding λ will be given as

$$\lambda_\nu = \begin{cases} \lambda_{kl} \text{ if } \langle \tilde{n}_k \rangle < \langle \tilde{n}_l \rangle, \\ \overset{*}{\lambda_{kl}} \text{ if } \langle \tilde{n}_k \rangle > \langle \tilde{n}_l \rangle, \\ 0 \quad \text{if } \langle \tilde{n}_k \rangle = \langle \tilde{n}_l \rangle. \end{cases} \tag{7.47}$$

The last choice is made for convenience since no such terms occur in Equation (7.46). Matrix elements $B_{\nu\nu'}$ and $C_{\nu\nu'}$ are defined

$$B_{\nu\nu'} = \delta_{kk'}\delta_{ll'}\,(\epsilon_k - \epsilon_l)\,[\langle \tilde{n}_l \rangle - \langle \tilde{n}_k \rangle] + [(kl\,|l'k') - (kk'|\,l'l)]\,[\langle \tilde{n}_l \rangle - \langle \tilde{n}_k \rangle]\,[\langle \tilde{n}_{l'} \rangle - \langle \tilde{n}_{k'} \rangle], \tag{7.48}$$

and

$$C_{\nu\nu'} = [(kl\,|k'l') - (kl'|\,k'l)]\,[\langle \tilde{n}_l \rangle - \langle \tilde{n}_k \rangle]\,[\langle \tilde{n}_{l'} \rangle - \langle \tilde{n}_{k'} \rangle]. \tag{7.49}$$

We also have

$$D_\nu = i\delta h_{kl}\,[\langle \tilde{n}_l \rangle - \langle \tilde{n}_k \rangle],\ \langle \tilde{n}_l \rangle < \langle \tilde{n}_k \rangle. \tag{7.50}$$

The formula for the second variation is now

$$\delta^2 \langle H \rangle = \langle \delta^2 H \rangle + \sum (D_\nu \lambda_\nu^* + D_\nu^* \lambda_\nu) + \tfrac{1}{2} \sum (B_{\nu\nu'} \lambda_\nu^* \lambda_{\nu'} + B_{\nu\nu'}^* \lambda_\nu \lambda_{\nu'}^* - C_{\nu\nu'}^* \lambda_\nu \lambda_{\nu'} - C_{\nu\nu'} \lambda_\nu^* \lambda_{\nu'}^*). \tag{7.51}$$

Most applications of the Hartree–Fock method lead to real spin orbitals so that $\mathbf{B}$ and $\mathbf{C}$ are real. We see then that in the absence of external perturbations, $\mathbf{D}=0$, the nature of the stationary point $\delta \langle H \rangle = 0$ is determined by the eigenvalues of the matrices $\mathbf{B} \pm \mathbf{C}$. If both these matrices are positive definite we have a minimum energy point since $\delta^2 \langle H \rangle$ then is non-negative. Such an extremum is the kind most often looked for, but under certain circumstances one is satisfied with stationary solutions with saddle-point character but having specified symmetry properties.

For a case with external time-independent perturbations the formula (7.51) can be used to calculate the second order perturbation energy in the Hartree–Fock approximation. One makes then a choice of λ_ν's such that $\delta^2 \langle H \rangle$ is stationary. The solution is obtained from the matrix equations

$$\begin{bmatrix} \mathbf{B} & -\mathbf{C} \\ -\mathbf{C} & \mathbf{B} \end{bmatrix} \begin{bmatrix} \boldsymbol{\lambda} \\ \boldsymbol{\lambda}^* \end{bmatrix} = - \begin{bmatrix} \mathbf{D} \\ \mathbf{D}^* \end{bmatrix}. \tag{7.52}$$

These equations are the so-called coupled Hartree–Fock perturbation equations. Various simplifications are used in the literature to avoid the complications by the electron interaction terms in the $\mathbf{B}$ and $\mathbf{C}$ matrics. The second order energy will now be, for $\mathbf{D} = \mathbf{D}_1 + i\mathbf{D}_2$ separated in real and imaginary parts,

$$\delta^2 \langle H \rangle = \langle \delta^2 H \rangle - \tfrac{1}{2} \mathbf{D}_1^\dagger\, [\mathbf{B} - \mathbf{C}]^{-1}\, \mathbf{D}_1 - \tfrac{1}{2} \mathbf{D}_2^\dagger\, [\mathbf{B} + \mathbf{C}]^{-1}\, \mathbf{D}_2. \tag{7.53}$$

The results of this kind of perturbation calculation is thus closely connected with the stability criterion discussed above. The eigenvalues of $\mathbf{B} \pm \mathbf{C}$ serve as energy differences in an expression corresponding to Rayleigh–Schrödinger perturbation theory,

$$\delta^2 \langle H \rangle = \langle \delta^2 H \rangle - \sum \rho_n \, |\langle n | \delta H | \, m \rangle|^2 \, (E_m - E_n)^{-1}. \tag{7.54}$$

It is clearly unsatisfactory to use a solution with saddle-point character for the calculation of ground state properties by means of second order perturbation theory of the Hartree–Fock equations.

The Particle-hole Propagator

Related to the stability question of Hartree–Fock solutions is the calculation of excitation energies and transition moments as already indicated at the end of the previous section. Green's functions of the type $\langle\langle a_r^\dagger a_s ; a_{s'}^\dagger a_{r'} \rangle\rangle_E$ are seen to have poles at energy differences between states of equal number of electrons. The corresponding residues are needed in the calculation of transition probabilities.

An approximation to the Green's function mentioned above can be obtained by the moment expansion method. The description of the procedure is given in terms of so-called particle-hole operators, which are defined, compare Equation (7.47), as

$$q_\nu^\dagger = \tilde{a}_k^\dagger \tilde{a}_l, \; \langle \tilde{n}_k \rangle < \langle \tilde{n}_l \rangle, \tag{7.55}$$

$$q_\nu = \tilde{a}_k^\dagger a_l, \; \langle \tilde{n}_k \rangle > \langle \tilde{n}_l \rangle, \tag{7.56}$$

whereas the operators $\tilde{a}_k^\dagger \tilde{a}_l$ with $\langle \tilde{n}_k \rangle = \langle \tilde{n}_l \rangle$ are not needed presently.

The Green's functions to be considered are collected in a matrix,

$$\mathbf{P}(E) = \begin{bmatrix} \langle\langle q_\nu ; q_{\nu'}^\dagger \rangle\rangle_E & \langle\langle q_\nu ; q_{\nu'} \rangle\rangle_E \\ \langle\langle q_\nu^\dagger ; q_{\nu'}^\dagger \rangle\rangle_E & \langle\langle q_\nu^\dagger ; q_{\nu'} \rangle\rangle_E \end{bmatrix} \tag{7.57}$$

to be understood as a blocked matrix with a representative element indicated in each block.

The equation of motion requires the calculation of the following quantities,

$$\langle [q_\nu, q_{\nu'}^\dagger] \rangle = -\langle [q_{\nu'}^\dagger, q_\nu] \rangle = \delta_{\nu\nu'} \, | \, \langle \tilde{n}_k \rangle - \langle \tilde{n}_l \rangle \, |, \tag{7.58}$$

$$\langle [q_\nu, q_{\nu'}] \rangle = \langle [q_\nu^\dagger, q_{\nu'}^\dagger] \rangle = 0, \tag{7.59}$$

$$\langle [\, [q_\nu, H], q_{\nu'}^\dagger] \rangle = \langle [\, [q_\nu^\dagger, H], q_{\nu'}] \rangle^* = B_{\nu\nu'}, \tag{7.60}$$

$$\langle [\, [q_\nu, H], q_{\nu'}] \rangle = \langle [\, [q_\nu^\dagger, H], q_{\nu'}^\dagger] \rangle^* = -C_{\nu\nu'}, \tag{7.61}$$

where the formation of average values again relies upon the choice of the ensemble (7.14).

We specialize now to the case of a standard Hartree–Fock solution with occupation numbers zero or unity which means that each $q^\dagger$ represents the elementary process of exciting an electron from an occupied orbital into an unoccupied one. Thus is obtained the concept of a particle-hole excitation. We see that the diagonal elements of Equation (7.58) are all unity and those of Equation (7.60) are

$$B_{\nu\nu} = \epsilon_k - \epsilon_l + (lk|kl) - (ll|kk), \tag{7.62}$$

which corresponds to energy difference between the Hartree–Fock ground state and the state generated from it by the operator $q_\nu^\dagger$.

The equation of motion for the matrix Green's function $\mathbf{P}(E)$ can now be iterated as we did in Equation (7.17) to obtain

$$\begin{aligned}\mathbf{P}(E) &= E^{-1}\begin{bmatrix}\mathbf{1} & \mathbf{0}\\ \mathbf{0} & -\mathbf{1}\end{bmatrix} + E^{-2}\begin{bmatrix}\mathbf{B} & -\mathbf{C}\\ -\mathbf{C}^* & \mathbf{B}^*\end{bmatrix} + \ldots\\ &= \begin{bmatrix}E\,\mathbf{1}-\mathbf{B} & -\mathbf{C}\\ -\mathbf{C}^* & -E\,\mathbf{1}-\mathbf{B}^*\end{bmatrix}^{-1}.\end{aligned} \tag{7.63}$$

The poles of $\mathbf{P}(E)$ occur as solutions of a non-hermitian eigenvalue problem and need not be real. The eigenvalue equation can be reduced, for real matrices, to finding the singularities of

$$[(\mathbf{B}+\mathbf{C})\,(\mathbf{B}-\mathbf{C}) - \mathrm{E}^2\mathbf{1}]^{-1}$$

which is hermitian only if $\mathbf{B}$ and $\mathbf{C}$ commute with one another.

The construction of an explicit representation of the Green's function can be achieved through a series of transformations whenever the Hartree–Fock solution corresponds to a minimum point such that $\mathbf{B} \pm \mathbf{C}$ are both positive definite. We consider the unitary matrix $\mathbf{U}$ which diagonalizes $\mathbf{B}+\mathbf{C}$:

$$(\mathbf{B}+\mathbf{C})\,\mathbf{U} = \mathbf{U}\mathbf{X}, \tag{7.64}$$

and use it for the transformation

$$\mathbf{X}^{\frac{1}{2}}\mathbf{U}^\dagger(\mathbf{B}-\mathbf{C})\,\mathbf{U}\mathbf{X}^{\frac{1}{2}} = \mathbf{W}^2, \tag{7.65}$$

where $\mathbf{X}^{\frac{1}{2}}$ is a diagonal matrix with positive elements, the square of which equals $\mathbf{X}$, and where the notation $\mathbf{W}^2$ is chosen to indicate the positive definiteness.

Let $\mathbf{V}$ be the unitary matrix which diagonalizes $\mathbf{W}^2$ to give positive eigenvalues denoted as the square of the positive diagonal matrix $\boldsymbol{\omega}$:

$$\mathbf{W}^2\,\mathbf{V} = \mathbf{V}\boldsymbol{\omega}^2. \tag{7.66}$$

Then we can define two intermediate matrices,

$$\mathbf{S} = \mathbf{U}\mathbf{X}^{-\frac{1}{2}}\,\mathbf{V}\boldsymbol{\omega}^{\frac{1}{2}}, \tag{7.67}$$

and

$$\mathbf{T} = \mathbf{U}\mathbf{X}^{\frac{1}{2}}\,\mathbf{V}\boldsymbol{\omega}^{\frac{1}{2}} = (\mathbf{S}^\dagger)^{-1}, \tag{7.68}$$

as well as their sum and difference,

$$\mathbf{Z}=\tfrac{1}{2}(\mathbf{S}+\mathbf{T}), \tag{7.69}$$

$$\mathbf{Y}=\tfrac{1}{2}(\mathbf{S}-\mathbf{T}), \tag{7.70}$$

which enter the final form of $\mathbf{P}(E)$:

$$\mathbf{P}(E)=\begin{bmatrix}\mathbf{Z} & \mathbf{Y}\\ \mathbf{Y} & \mathbf{Z}\end{bmatrix}\begin{bmatrix}(E\mathbf{1}-\boldsymbol{\omega})^{-1} & \mathbf{0}\\ \mathbf{0} & -(E\mathbf{1}+\boldsymbol{\omega})^{-1}\end{bmatrix}\begin{bmatrix}\mathbf{Z}^{\dagger} & \mathbf{Y}^{\dagger}\\ \mathbf{Y}^{\dagger} & \mathbf{Z}^{\dagger}\end{bmatrix}. \tag{7.71}$$

By comparison of this formula with the general spectral resolution (5.14) we identify the eigenvalues as

$$\omega_j=E_j-E_0, \tag{7.72}$$

and the matrix elements as

$$Z_{\nu j}=\langle 0\,|q_\nu|\,j\rangle, \tag{7.73}$$

and

$$Y_{\nu j}=\langle 0\,|q_\nu^{\dagger}|\,j\rangle. \tag{7.74}$$

This indicates immediately that the ground state $|\,0\rangle$ cannot generally be interpreted as the Hartree–Fock state since $Y_{\nu j}$ should equal zero which only holds when $\mathbf{C}$ vanishes. The approximation is then equivalent to a configuration interaction calculation among so-called mono-excited states.

The moment expansion method that we have used above for the particle-hole propagators is equivalent to several other derivations of the same equations. Thus the equations are known as the time-dependent Hartree–Fock equations or the random phase approximation. They are not self-consistent in the fashion of the ordinary Hartree–Fock equations for the electron propagator since the expectation values

$$\langle q_\nu^{\dagger}\, q_{\nu'}\rangle=\sum Y_{\nu j}Y_{\nu' j}^{*}$$

are not zero as for the Hartree–Fock state. Average values calculated from the time-dependent Hartree–Fock approximation to the particle-hole propagator does not satisfy certain symmetry conditions following from the anti-commutation rules for the field operator. For instance, it cannot be assured that the following identity holds:

$$\langle a_k^{\dagger}a_l a_{k'}^{\dagger}a_{l'}\rangle=-\langle a_k^{\dagger}a_{l'}a_{k'}^{\dagger}a_l\rangle+\delta_{k'l'}\langle a_k^{\dagger}a_l\rangle+\delta_{k'l}\,\langle a_k^{\dagger}a_{l'}\rangle. \tag{7.75}$$

Thus it is a doubtful procedure to use the time-dependent Hartree–Fock method to calculate an estimate to the total energy.

A form of the method which neglects exchange was studied by Lindhard for an electron gas. It has been highly successful for the description of the dynamical properties and the total energy for this model system in the high density limit.

Problem 1

Consider the three parameter hamiltonian

$$H=\alpha(a_1^\dagger a_1+a_2^\dagger a_2+a_3^\dagger a_3+a_4^\dagger a_4)$$
$$+\beta(a_1^\dagger a_3+a_2^\dagger a_4+a_3^\dagger a_1+a_4^\dagger a_2)+\gamma(a_1^\dagger a_2^\dagger a_2 a_1+a_3^\dagger a_4^\dagger a_4 a_3),$$

which might represent the truncation of the full hamiltonian to a manifold of states generated by two valence orbitals (four spin orbitals), each on one of two equivalent atoms. The bond parameter is β and all but one center electron interaction integrals have been discarded. Solve the Hartree-Fock problem for the case of two electrons in the normal state and find the condition of stability in terms of the ratio β/γ. [Stable if $0<\gamma<|\beta|$.]

Problem 2

Find the excitation energies for the system studied in the previous problem from the time-dependent Hartree–Fock approximation. [$\omega=[4\beta(\beta\pm\gamma)]^{\frac{1}{2}}$.]

Notes and Bibliography

The self-consistent field method was introduced in many-electron quantum mechanics by D. R. Hartree in *Proc. Camb. Phil. Soc.* **24**, 105 (1928), and his ideas spurred an intensive development, particularly by J. C. Slater and V. Fock. The latter gave a derivation of the basic equations in 1932 using second quantization, see *Z. Physik* **75**, 622. P. A. M. Dirac formulated the time-dependent version in a paper in *Proc. Camb. Phil. Soc.* **26**, 376, from 1930. J. Lindhard gave in 1954 an explicit solution for an electron gas and demonstrated its application to plasma oscillations in metals, *Kgl. Danske Videnskabers Selskab, Mat.-Fys. Medd.* **28**, No. 8. The connection of these equations to the stability of Hartree–Fock states was emphasized in 1960 by D. J. Thouless *Nucl. Phys.* **21**, 225.

CHAPTER 8

The Atomic Central Field Problem

We are going to explore in this chapter some of the consequences of the spherical symmetry of the hamiltonian for an atomic system in the structure of the Green's function. A number of results concerning tensor operators are invoked and we assume familiarity with $3j$ and $6j$ symbols.

Electron Propagator

It was shown in Chapter 6 that the electron propagator for a spherically symmetric potential could be expressed as a convolution of an angular and a radial factor Green's function. Such a separation is also possible for a many-electron system when the ensemble which is used in the definition of Green's functions is invariant under rotations around the atomic nucleus. The many-electron states labeled n, m in Equations (5.9) and (5.10) will in the present case be characterized by explicit labels of total orbital and spin angular momentum quantum numbers, assuming Russell–Saunders coupling,

$$n \to \gamma L M_L S M_S, \; m \to \gamma' L' M_L' S' M_S',$$

where γ, γ' symbolize other possible quantum numbers. The density operator has now the diagonal elements

$$\rho_n = \rho_{\gamma L M_L S M_S} = w(\gamma L S)/[(2L+1)(2S+1)], \tag{8.1}$$

when the ensemble density operator commutes with the operators of orbital and spin angular momentum as given in Chapter 4.

The electron field operator will be expanded in a basis set adapted to spherical symmetry so that each spin orbital $u_s(\xi)$ is a product of a radial function, a spherical harmonic, and a spin function

$$u_s(\xi) = R_{nl}(r) Y_{lm}(\theta, \phi) \delta_{\nu\zeta}. \tag{8.2}$$

The spin function is here given as a Kronecker δ where the spin variable ζ and the spin quantum number ν both can assume values in the set $\pm\frac{1}{2}$.

This corresponds to the definitions, for the conventional spin functions,

$$\alpha(\zeta) = \delta_{\frac{1}{2}\zeta} \text{ and } \beta(\zeta) = \delta_{-\frac{1}{2}\zeta}. \tag{8.3}$$

The spherical harmonics are defined with phase conventions such that

$$Y_{ll}(\theta, \phi) = [(2l+1)!/4\pi]^{\frac{1}{2}}(l!)^{-1}\, [-\tfrac{1}{2}\sin\theta e^{i\phi}]^l, \tag{8.4}$$

and

$$Y_{lm}(\theta, \phi) = [(l+m)!/(2l)!\,(l-m)!]^{\frac{1}{2}}\, [e^{-i\phi}\,(-\partial/\partial\theta + i\cot\theta\partial/\partial\phi)]^{l-m}\, Y_{ll}(\theta, \phi) \tag{8.5}$$

which is in accord with the notation of Condon and Shortley in their book "The Theory of Atomic Spectra".

The expansion for the electron field operator is

$$\psi(\xi) = \sum_{nlm\nu} R_{nl}(r)\, Y_{lm}(\theta, \phi)\delta_{\nu\xi} a_{nlm\nu}, \tag{8.6}$$

and we obtain for operators of total orbital and spin angular momentum respectively,

$$\begin{aligned} L_+ &= L_x + iL_y = \sum_{nlm\nu}[(l-m)\,(l+m+1)]^{\frac{1}{2}}\, a^\dagger_{nlm+1\nu} a_{nlm\nu}, \\ L_- &= L_x - iL_y = \sum_{nlm\nu}[(l+m)\,(l-m+1)]^{\frac{1}{2}}\, a^\dagger_{nlm-1\nu}\, a_{nlm\nu}, \\ L_z &= \sum_{nlm\nu} m\, a^\dagger_{nlm\nu}\, a_{nlm\nu}, \end{aligned} \tag{8.7}$$

$$\begin{aligned} S_+ &= S_x + iS_y = \sum_{nlm} a^\dagger_{nlm\frac{1}{2}}\, a_{nlm-\frac{1}{2}}, \\ S_- &= S_x - iS_y = \sum_{nlm} a^\dagger_{nlm-\frac{1}{2}}\, a_{nlm\frac{1}{2}}, \\ S_z &= \tfrac{1}{2}\sum_{nlm}\, [a^\dagger_{nlm\frac{1}{2}}\, a_{nlm\frac{1}{2}} - a^\dagger_{nlm-\frac{1}{2}}\, a_{nlm-\frac{1}{2}}], \end{aligned} \tag{8.8}$$

when the radial functions $R_{nl}(r)$ form an orthonormal set,

$$\int R^*_{nl}(r) R_{n'l}(r) r^2 dr = \delta_{nn'}. \tag{8.9}$$

The creation operator $a^\dagger_{nlm\nu}$ is a tensor operator of rank l with respect to the orbital angular momentum and of rank $\frac{1}{2}$ with respect to the spin angular momentum, for instance

$$[L_+, a^\dagger_{nlm\nu}] = [(l-m)\,(l+m+1)]^{\frac{1}{2}}\, a^\dagger_{nlm+1\nu}, \tag{8.10}$$

and

$$[S_+, a^\dagger_{nlm\nu}] = [(\tfrac{1}{2}-\nu)\,(\tfrac{1}{2}+\nu+1)]^{\frac{1}{2}}\, a^\dagger_{nlm\nu+1} = \delta_{-\frac{1}{2}\nu}\, a^\dagger_{nlm\frac{1}{2}}. \tag{8.11}$$

Similar formulae hold for $a_{nlm\nu}$, but one must realize that $(-)^{l-m+\frac{1}{2}-\nu} a_{nlm\nu}$ transforms as $a^\dagger_{nl-m-\nu}$.

Matrix elements of the field operator $\psi^\dagger(\xi)$ can now be calculated in terms of reduced matrix elements of the tensor operators $a^\dagger_{nlm\nu}$ using the Wigner–Eckart theorem. We obtain

$$\begin{aligned} &\langle \gamma L M_L S M_S\, |\psi^\dagger(\xi)|\, \gamma' L' M'_L S' M'_S \rangle \\ &= \sum_{nlm\nu} R^*_{nl}(r)\, Y^*_{lm}(\theta, \phi)\delta_{\nu\xi}(-)^g \begin{pmatrix} L & l & L' \\ -M_L & m & M'_L \end{pmatrix} \begin{pmatrix} S & \frac{1}{2} & S' \\ -M_S & \nu & M'_S \end{pmatrix} \\ &\qquad \times \langle \gamma L S\, \|a^\dagger_{nl}\|\, \gamma' L' S' \rangle, \end{aligned} \tag{8.12}$$

with

$$g = L - M_L + S - M_S.$$

The reduced matrix element and the radial function in Equation (8.12) are the only factors that depend upon the label n and one might define a more general reduced matrix element by the relation

$$\langle \gamma LS \| R_l^\dagger(r) \| \gamma' L' S' \rangle = \sum_n \overset{*}{R}_{nl}(r) \langle \gamma LS \| a_{nl}^\dagger \| \gamma' L' S' \rangle, \qquad (8.13)$$

which is independent of the particular choice of radial functions. It is a consequence of the transformation properties of the annihilation operator $\psi(\xi)$ and the individual $a_{nlm\nu}$ that the corresponding reduced matrix elements are defined as

$$\begin{aligned} \langle \gamma' L' S' \| R_l(r) \| \gamma LS \rangle &= \sum_n R_{nl}(r) \langle \gamma' L' S' \| a_{nl} \| \gamma LS \rangle \\ &= (-)^{L'+l-L+S'+\frac{1}{2}-S} \langle \gamma LS \| R_l^\dagger(r) \| \gamma' L' S' \rangle^*. \end{aligned} \qquad (8.14)$$

At present we have not considered other symmetry properties than orbital and spin rotations but it is obvious that in Russell–Saunders coupling we can also find that parity considerations will limit the possible l values further so that only even (odd) values are allowed when γ and γ' have equal (different) parity quantum numbers.

The form (8.12) allows us to perform partial summations in the spectral representation of the electron propagator. Thus we have

$$\begin{aligned} &\sum_{M_L M_L' M_S M_S'} \langle \gamma' L' M_L' S' M_S' | \psi(\xi) | \gamma L M_L S M_S \rangle \\ &\qquad\qquad \times \langle \gamma L M_L S M_S | \psi^\dagger(\xi') | \gamma' L' M_L' S' M_S' \rangle \\ &= (\delta_{\zeta\zeta'}/8\pi) \sum_l P_l(\cos\Theta) \langle \gamma LS \| R_l^\dagger(r) \| \gamma' L' S' \rangle^* \langle \gamma LS \| R_l^\dagger(r') \| \gamma' L' S' \rangle, \end{aligned} \qquad (8.15)$$

with

$$\begin{aligned} \cos\Theta &= \sin\theta \sin\theta' \cos(\phi-\phi') + \cos\theta\cos\theta' \\ &= \bar{r}.\bar{r}'/rr', \end{aligned} \qquad (8.16)$$

where the orthogonality of the $3j$-symbols and the spherical harmonic addition theorem have been utilized. It follows that the electron propagator is now obtained as

$$G(\xi, \xi'; E) = \delta_{\zeta\zeta'} \sum_l \frac{2l+1}{4\pi} P_l(\cos\Theta)\, G_{l(l+1)}(r, r'; E)/rr', \qquad (8.17)$$

where we have used the result of the convolution integral (6.11). Identification of Equation (8.17) and the spectral form (5.12) expressed with the notation (8.1) and the result (8.15) gives

$$\begin{aligned} &G_{l(l+1)}(r, r'; E) \\ &= \sum [rr'/(4l+2)] \langle \gamma LS \| R_l^\dagger(r) \| \gamma' L' S' \rangle^* \langle \gamma LS \| R_l^\dagger(r') \| \gamma' L' S' \rangle \\ &\qquad \times \left[\frac{w(\gamma LS)/(2L+1)(2S+1)}{E - E(\gamma' L' S') + E(\gamma LS) - i\eta} + \frac{w(\gamma' L' S')/(2L'+1)(2S'+1)}{E - E(\gamma' L' S') + E(\gamma LS) + i\eta} \right]. \end{aligned} \qquad (8.18)$$

We wish to compare the expression (8.18) with the spectral form of the solution to the radial Green's function equation (6.14),

$$G_{l(l+1)}(r, r'; E) = \sum_k u_k(r) u_k^*(r')/(E - \epsilon_k), \qquad (8.19)$$

with orthonormal functions $u_k(r)$:

$$\int u_k^*(r) u_{k'}(r) dr = \delta_{kk'}. \qquad (8.20)$$

The corresponding amplitudes in the case of a system of interacting electrons are proportional to $\langle \gamma' L' S' \| R_l(r) \| \gamma L S \rangle$ but they do not generally form a linearly independent set. The simplest theory of the spectra of atoms describes the many-electron states $|\gamma L M_L S M_S\rangle$ in terms of configurations, where a prescribed number of spin orbitals from each subshell (nl) in the basis set are occupied. That is, these states are eigenstates of the subshell number operators

$$N_{nl} = \sum_{m\nu} a^\dagger_{nlm\nu} a_{nlm\nu}. \qquad (8.21)$$

Let us consider an atomic term (γLS), corresponding to eigenvalues $N_{nl}(\gamma)$ for the operators N_{nl} respectively. This term can be thought of as arising from another term, $(\gamma' L' S')$, by adding one electron in the subshell (nl). Thus we have

$$N_{nl}(\gamma) = N_{nl}(\gamma') + 1 \qquad (8.22)$$

for this subshell and equal eigenvalues for all other subshells. Only one reduced matrix element $\langle \gamma LS \| a^\dagger_{n'l'} \| \gamma' L' S' \rangle$ is then different from zero and it serves to define the coefficient of fractional parentage through the relation

$$\langle \gamma LS \| a^\dagger_{nl} \| \gamma' L' S' \rangle = (-)^g \, [g(2L+1)(2S+1)]^{\frac{1}{2}} \, (\gamma LS\{|\gamma' L' S'), \qquad (8.23)$$
$$g = N_{nl}(\gamma).$$

The following orthonormality condition,

$$\sum_{\gamma' L' S'} (\gamma LS\{|\gamma' L' S')\, (\gamma'' LS\}|\gamma' L' S')^* = \delta_{\gamma\gamma''}, \qquad (8.24)$$

is satisfied by the coefficients of fractional parentage and is easily shown from the definitions (8.23), (8.12) and (8.21).

The orbital configuration approximation is seen to give, in Equation (8.18), an expansion similar to the one in Equation (8.19) in that the radial amplitudes now form an orthonormal set, but the factor $(E - \epsilon_{nl})^{-1}$ is replaced by a sum of terms, $\sum_j c_j (E - \epsilon_j)^{-1}$, each pole of which corresponds to a transition from a term (γLS) to another.

A particularly simple example of an atomic propagator obtains in the pure state case (Equation 5.13) for a half-filled subshell. Then all terms have the weight zero except one which has $L = 0$ and $S = l + \frac{1}{2}$. These two quantum numbers specify the term uniquely. The Pauli exclusion principle then gives the following selection rule for the coefficients of fractional parentage.

$$(0, l+\tfrac{1}{2}\{|\gamma LS)=\delta_{Ll}\delta_{Sl}\,(0, l+\tfrac{1}{2}\{|l, l)=\delta_{Ll}\delta_{Sl}, \tag{8.25}$$

and

$$(\gamma LS\{|\,0, l+\tfrac{1}{2})=\delta_{Ll}\delta_{Sl}(2l+1)^{-\frac{1}{2}}. \tag{8.26}$$

The final expression for the electron propagator is

$$\langle\langle a_{m\nu}; a^{\dagger}_{m'\nu'}\rangle\rangle_E=\delta_{mm'}\delta_{\nu\nu'}\left(\frac{\tfrac{1}{2}}{E-E_1+i\eta}+\frac{\tfrac{1}{2}}{E-E_2-i\eta}\right) \tag{8.27}$$

with

$$E_1=E((l)^{2l+1}, 0, l+\tfrac{1}{2})-E((l)^{2l+2}, l, l), \tag{8.28}$$

and

$$E_2=E((l)^{2l}, l, l)-E((l)^{2l+1}, 0, l+\tfrac{1}{2}). \tag{8.29}$$

This simple result has a direct relation to the establishment of parametrized molecular models as will be seen in Chapter 9.

We can express the electron propagator in terms of the spectral weights $A(\xi, \xi'; s)$ and according to Equation (3.28) we have

$$G(\xi, \xi'; E)=\sum_s \frac{A(\xi, \xi'; s)}{E-\epsilon_s \mp i\eta}, \tag{8.30}$$

where we have defined the energies,

$$\epsilon_s=E_s\,(N+1)-E_0(N) \tag{8.31}$$

and

$$\epsilon_s=E_0(N)-E_s(N-1), \tag{8.32}$$

of a spectrum which stretches from infinitely negative to infinitely positive values and which has a gap at an energy equal to the negative of the electronegativity of the atom on the Mulliken scale; the gap arises since

$$\min E_s(N+1)-E_0(N)> \max E_0(N)-E_s(N-1). \tag{8.33}$$

This is equivalent to stating that the ionization potential of the $(N+1)$-electron ion is smaller than the one of the N-electron atom. The electronegativity as defined by Mulliken is the average of the corresponding energies and can be thought of as a representative value for the quantity $-\partial E(N)/\partial N$ or the negative of the chemical potential. The spectral weight is readily obtained from Equation (5.12) as

$$A(\xi, \xi'; s)=\sum_{nm}^{\varepsilon_s=E_m-E_n} \langle n\,|\psi(\xi)|\,m\rangle\,\langle m\,|\psi^{\dagger}(\xi')|\,n\rangle\,[\rho_n+\rho_m], \tag{8.34}$$

where the energy selection rule seen from Equation (3.25) limits the summation to a restricted set of states. As we have seen before this also influences the choice of imaginary infinitesimals in Equation (8.30), such that poles of the Green's function lie below the real axis in the E-plane when Equation (8.31) applies and immediately above when Equation (8.32) holds.

The energy spectrum $[\epsilon_s]$ and the spectral weights $[A(\xi, \xi'; s)]$ are directly useful for the determination of atomic properties. Let us define the energy parameter α as the midpoint of the energy gap discussed above, or equal to the negative of the electronegativity as defined by Mulliken. The parameter then is expressed as

$$\alpha = \tfrac{1}{2} \max [E_0(N) - E_s(N-1)] + \tfrac{1}{2} \min [E_s(N+1) - E_0(N)]$$
$$= -\tfrac{1}{2}[I(N) + I(N+1)], \tag{8.35}$$

where $I(N)$ is the energy required to remove an electron from the N-particle system. The one-particle reduced density matrix as given in Equation (5.21) can be expressed as

$$\langle \psi^{\dagger}(\xi')\, \psi(\xi) \rangle = \sum_{\varepsilon_s < \alpha} A(\xi, \xi'; s), \tag{8.36}$$

and from Equation (4.8) we obtain directly the sumrule

$$N = \int \langle \psi^{\dagger}(\xi)\, \psi(\xi) \rangle \, d\xi = \sum_{\varepsilon_s < \alpha} \int A(\xi, \xi; s) d\xi. \tag{8.37}$$

Further, the sumrule

$$\sum_{s} A(\xi, \xi'; s) = \langle \psi(\xi) \psi^{\dagger}(\xi') + \psi^{\dagger}(\xi') \psi(\xi) \rangle = \delta(\xi - \xi') \tag{8.38}$$

is readily verified from Equation (3.25).

The transition amplitude, $\langle \gamma NLM_L SM_S \,|\psi^{\dagger}(\xi)|\, \gamma' N' L' M'_L S' M'_S \rangle$ may be interpreted as the complex conjugate [see Equation (8.14)] of the wave function for an electron that has been added to the state $|\gamma' N' L' M'_L S' M'_S\rangle$ of N' electrons to produce the state $|\gamma NLM_L SM_S\rangle$ of N electrons. The angular part of this wave function will generally be very restricted as to its functional form, while there is no general symmetry argument which will determine the radial part. Approximate calculations of these radial amplitudes, for the hydrogen, carbon and nitrogen atoms are presented in Appendix I. The contribution of such a matrix element of the electron field operator, $\psi^{\dagger}(\xi)$, between two Russell–Saunders states, to the spectral weight matrix is obtained from Equations (8.15) and (8.18) as

$$A(\xi, \xi'; s) = (\delta_{\zeta\zeta'}/8\pi) \sum_{l} P_l(\cos \Theta) \langle \gamma NLS || R_l^{\dagger}(r) || \gamma' N' L' S' \rangle^*$$
$$\times \langle \gamma NLS || R_l^{\dagger}(r') || \gamma' N' L' S' \rangle$$
$$\times [w(\gamma NLS)/(2L+1)(2S+1) + w(\gamma' N' L' S')/(2L'+1)(2S'+1)], \tag{8.39}$$

and from Equation (8.36) it follows that the amplitudes

$$\langle \gamma NLS || R_l^\dagger(r) || \gamma' N'L'S' \rangle$$

will, for $\epsilon_s < \alpha$, determine the charge distribution in the atom just as the occupied orbitals do in the Hartree–Fock approximation.

Hartree–Fock Equations

We will repeat in this section some of the results from Chapter 7 with particular reference to spherical symmetry. The spin orbital basis is defined by Equation (7.2), and the density operator is introduced in Equation (7.26). This density operator is invariant under rotations of the coordinate system only if the quantities $\langle n_s \rangle = \langle n_{(nlm\nu)} \rangle$ are independent of m and ν. We denote the average number of electrons in the subshell by $\langle N_{nl} \rangle = q(nl)$ and obtain

$$\langle n_{(nlm\nu)} \rangle = q(nl)/(4l+2). \tag{8.40}$$

The average value of the total hamiltonian is then obtained from Equations (7.32) and (7.33) as

$$\langle H \rangle = \sum_s h_{ss} \langle n_s \rangle + \tfrac{1}{2} \sum [(ss\,|tt) - (st|\,ts)] \langle n_s \rangle \langle n_t \rangle. \tag{8.41}$$

We introduce the notation $h_{(nlm\nu)} = I(nl)$, since these integrals do not depend on m and ν. The electron interaction integrals are in detail expressed in terms of $3j$-symbols and Slater integrals R^k as

$$\begin{aligned}
&(n_1 l_1 m_1 \nu_1, n_2 l_2 m_2 \nu_2\, | n_3 l_3 m_3 \nu_3, n_4 l_4 m_4 \nu_4) \\
&\quad = \delta_{\nu_1\nu_2} \delta_{\nu_3\nu_4} (-)^{-m_1-m_4} \\
&\quad \times \sum_{k\mu} \begin{pmatrix} l_1 & k & l_2 \\ -m_1 & \mu & m_2 \end{pmatrix} \begin{pmatrix} l_4 & k & l_3 \\ -m_4 & \mu & m_3 \end{pmatrix} \begin{pmatrix} l_1 & k & l_2 \\ 0 & 0 & 0 \end{pmatrix} \begin{pmatrix} l_4 & k & l_3 \\ 0 & 0 & 0 \end{pmatrix} \\
&\quad \times [(2l_1+1)(2l_2+1)(2l_3+1)(2l_4+1)]^{\frac{1}{2}} R^k(n_1 l_1 n_3 l_3, n_2 l_2 n_4 l_4).
\end{aligned} \tag{8.42}$$

The only integrals occurring will be

$$F^k(nl, n'l') = \int_0^\infty \int_0^\infty |R_{nl}(r_1)|^2 (r_<^k / r_>^{k+1}) |R_{n'l'}(r_2)|^2 r_1^2 r_2^2 \, dr_1 \, dr_2, \tag{8.43}$$

when in $R^k(n_1 l_1 n_3 l_3, n_2 l_2 n_4 l_4)$ we have that

$$(n_1 l_1) = (n_2\, l_2) = (nl),\ (n_3\, l_3) = (n_4\, l_4) = (n'\, l'),$$

and

$$G^k(nl, n'l') = \int_0^\infty \int_0^\infty R_{nl}^*(r_1) R_{n'l'}(r_1) (r_<^k / r_>^{k+1}) R_{nl}^*(r_2) R_{n'l'}(r_2) r_1^2 r_2^2 \, dr_1 \, dr_2, \tag{8.44}$$

when

$$(n_1 l_1) = (n_4 l_4) = (nl),\ (n_3 l_3) = (n_2 l_2) = (n'l').$$

In these expressions $r_<$ and $r_>$ denote the smaller and greater of r_1 and r_2 respectively.

The summation over magnetic and spin quantum numbers in the electron interaction terms of the energy functional, $\langle H\rangle$, leads to the definition of average interaction parameters, which we denote $V(nl, n'l')$ and define as

$$V(nl, n'l') = F^0(nl, n'l') - \tfrac{1}{2}\sum_k \begin{pmatrix} l & k & l' \\ 0 & 0 & 0 \end{pmatrix}^2 G^k(nl, n'l'). \tag{8.45}$$

The final formula for the energy functional $\langle H\rangle$ then is

$$\langle H\rangle = \sum_{nl} I(nl)q(nl) + \tfrac{1}{2}\sum_{nln'l'} q(nl)q(n'l')\, V(nl, n'l'). \tag{8.46}$$

We introduce the radial functions $P_{nl}(r) = r\,R_{nl}(r)$ and define the potential functions

$$U_k(nl, n'l'; r) = r^{-k-1}\int_0^r dr'(r')^k P_{nl}(r')P_{n'l'}(r') + r^k\int_r^\infty dr'(r')^{-k-1}P_{nl}(r')P_{n'l'}(r'), \tag{8.47}$$

and form the functional derivative of $\langle H\rangle$ with respect to $P_{nl}(r)$ as

$$\begin{aligned}\delta\langle H\rangle/\delta P_{nl}(r) = {} & q(nl)\,[-\tfrac{1}{2}P''_{nl}(r) + \tfrac{1}{2}l\,(l+1)\,r^{-2}P_{nl}(r) - (Z/r)P_{nl}(r) \\ & + \sum_{n'l'} q(n'l')\, U_0(n'l', n'l'; r)P_{nl}(r) \\ & - \tfrac{1}{2}\sum_{n'l'k} q(n'l') \begin{pmatrix} l & k & l' \\ 0 & 0 & 0 \end{pmatrix}^2 U_k\,(n'l', nl; r)P_{n'l'}(r)] \\ = {} & q(nl)\, F_l P_{nl}\,(r),\end{aligned} \tag{8.48}$$

where we have used atomic units and denoted the integro-differential operator by F_l. It is the effective operator of the Hartree–Fock equations determining the radial factors $P_{nl}(r)$. The orthonormality constraints on the orbitals bring in lagrangian multipliers $\epsilon(nl)$ and $\epsilon(nl, n'l')$ such that the vanishing of the variation of the energy functional, with the orthonormality constraints included, yields the equations

$$F_l P_{nl}(r) = \epsilon(nl)P_{nl}(r) + \sum_{n'\neq n} \epsilon(nl, n'l)P_{n'l}(r). \tag{8.49}$$

The off-diagonal lagrangian multipliers $\epsilon(nl, n'l)$ can be eliminated by choosing the $P_{nl}(r)$'s as eigenfunctions of F_l. There is no need for off-diagonal multipliers to ensure the orthogonality of the radial factors since the potential is the same for all orbitals; as is evident from the form of $\langle H\rangle$ in Equation (8.46) and the resulting effective operator F_l defined in Equation (8.48).

We will now assume that the orbitals have been determined such that variations of $\langle H\rangle$, with respect to radial factors, $P_{nl}(r)$, vanish and study the partial derivatives with respect to occupation numbers $q(nl)$:

$$\begin{aligned}\partial\langle H\rangle/\partial q(nl) &= I(nl)+\sum_{n'l'} q(n'l')\,V(nl,n'l') \\ &= \epsilon(nl).\end{aligned} \tag{8.50}$$

The parameters $\epsilon(nl)$ which occur as eigenvalues for the radial equations are the orbital energies and are thus obtained as derivatives of the total energy. Minimization of the expectation value of the energy while keeping the expectation value of the number operator constant leads to the condition

$$\partial\,[\langle H\rangle-\mu\,\langle N_{\mathrm{op}}\rangle]/\partial q(nl)\leq 0 \tag{8.51}$$

while

$$0\leq q(nl)\leq 4l+2. \tag{8.52}$$

The lagrangian multiplier μ appears similarly as the chemical potential in statistical mechanics and we see that it has a value such that it separates occupied from unoccupied orbital energy levels. The minimum total energy $\langle H\rangle$ is obtained when

$$\begin{aligned} q(nl)&=4l+2, && \epsilon(nl)<\mu, \\ q(nl)&=0, && \epsilon(nl)>\mu, \\ 0\leq q(nl)&\leq 4l+2, && \epsilon(nl)=\mu, \end{aligned} \tag{8.53}$$

and

$$\Sigma_{nl}q(nl)=\langle N_{\mathrm{op}}\rangle=N. \tag{8.54}$$

These relations are seen to lead to several open subshells only in the case when the corresponding orbital energies are degenerate.

TABLE 1. Comparison of Central Field Calculations for Fe, Co, and Ni Atoms. Energies in Hartrees

	$\langle H\rangle$	$\varepsilon\,(3d)$	$\varepsilon\,(4s)$
Fe $3d^6 4s^2$			
I	−1261·304	−0·3474	−0·2718
II	−1262·291	−0·6079	−0·2601
Co $3d^7\,4s^2$			
I	−1380·388	−0·4472	−0·2773
II	−1381·309	−0·6531	−0·2686
Ni $3d^8\,4s^2$			
I	−1506·069	−0·5531	−0·2825
II	−1506·816	−0·6971	−0·2768

I: Derived from Equation (8.46). II: Derived from Slater's average of configuration energy functional (from J. B. Mann).

Complex Spectra

The hamiltonian in Equation (7.32) can be divided into an unperturbed part consisting of the single particle terms chosen as the Fock operator of Equation (7.21) in diagonal form, and a perturbation consisting of a modified interaction term. Thus we find

$$H=\sum \epsilon_r n_r+\sum (rs|r's')\,[a_r^\dagger a_{r'}^\dagger a_{s'}a_s-a_r^\dagger \langle a_{r'}^\dagger a_{s'}\rangle a_s-a_{r'}^\dagger \langle a_r^\dagger a_s\rangle a_{s'} \\ +a_r^\dagger \langle a_{r'}^\dagger a_s\rangle a_{s'}+a_{r'}^\dagger \langle a_r^\dagger a_{s'}\rangle a_s]. \quad (8.55)$$

The perturbation term plays an important role for open shell atoms as has been shown by Slater. He used first order perturbation theory and Russell–Saunders states $|\gamma LM_L SM_S\rangle$ to calculate multiplet energies. Exactly the same treatment can be given here although our Fock operator, in the case of

TABLE 2. Multiplet Separations from Central Field Calculations and Experiment for Some Atoms in Units of cm^{-1}

Atom	Term	Calculated	Observed
	^3P	0	0
CI $2p^2$	^1D	10211	10164
	^1S*	17706	21619
	$^4S^0$	0	0
NI $2p^3$	$^2D^0$	20202	19200
	$^2P^0$*	26917	28840
	^3P	0	0
OI $2p^4$	^1D	16556	15790
	^1S*	29517	33700
	^3F	0	0
TiIII $3d^2$	^1D	9141	8231
	^3P	11169	10419
	^1G	14354	14156
	^5D	0	0
	^3P	22500	18551
	^3H	20200	19173
	^3F	23200	20411
FeI $3d^6\,4s^2$	^3G	25600	23636
	^1I	30300	28910
	^3D	33200	28953
	^1G	31300	29396
	^1D	30900	34234

* These terms have been corrected for configuration interaction $s^2p^n \longleftrightarrow p^{n+2}$.

open shells, is different from the one derived by the conventional variational treatment, since we have employed a density operator, which leads to averages involving configurations with different numbers of electrons. We show in Table 2 some multiplet energy differences calculated with the central field, described here and with the expressions for the multiplet energies given by Slater.

The agreement between theory and experiment is quite satisfactory and is better than that obtained with conventional central fields. It should be observed that our particular way of averaging leads to residual self energy terms in $\langle H\rangle$. For example, one electron in an s-subshell corresponds to $\langle n_s\rangle=\frac{1}{2}$ and the density operator gives rise to a binomial distribution over the configurations s^0, s^1 and s^2 with weights $w(s^0)=1/4$, $q(s^1)=2/4$ and $w(s^2)=1/4$. The expectation value of the hamiltonian is then

$$\langle H\rangle=\sum_{k=0}^{2} q(s^k)\, E_{\text{av}}(s^k), \tag{8.56}$$

where $E_{\text{av}}(s^k)$ is the Slater average of the configuration. For the neutral lithium atom we have

$$\langle H\rangle=2I(1s)+I(2s)+F^0(1s, 1s)+2F^0(1s, 2s)-G^0(1s, 2s)+\tfrac{1}{4}F^0(2s, 2s) \tag{8.57}$$

where the last term represents a certain amount of self-interaction in the open subshell. Positive contributions, like this, will result in somewhat higher total energies $\langle H\rangle$ than what is obtained from averaging with a fixed number of electrons. There is no physical significance in this fact alone, but the importance lies in the novel central field obtained and the spin orbitals it defines. Resulting multiplet splittings and one-electron energies are quite reasonable.

Single Subshell Approximation and Pair Creation Operators

We limit our discussion in this section to a single open subshell and consider only the perturbation within this shell. The summations in the perturbation term in Equation (8.55) then run only over electron states $(nlm\nu)$ of the open subshell.

The electron repulsion integrals will now be expressed as

$$\begin{aligned}
&(nlm_1\nu_1 nlm_2\nu_2 \mid nlm_3\nu_3 nlm_4\nu_4)\\
&\quad=\delta_{\nu_1\nu_2}\delta_{\nu_3\nu_4}\sum_{k\mu}F^k(nl, nl)\,(2l+1)^2\\
&\qquad\times\begin{pmatrix} l & k & l\\ 0 & 0 & 0\end{pmatrix}^2 \begin{pmatrix} l & k & l\\ -m_1 & \mu & m_2\end{pmatrix} \begin{pmatrix} l & k & l\\ -m_3 & -\mu & m_4\end{pmatrix}(-)^{-m_1-m_3-\mu} \qquad (8.58)\\
&\quad=\delta_{\nu_1\nu_2}\delta_{\nu_3\nu_4}\sum_{kLM}F^k\,(2l+1)^2\begin{pmatrix} l & k & l\\ 0 & 0 & 0\end{pmatrix}^2\begin{Bmatrix} l & l & k\\ l & l & L\end{Bmatrix}\\
&\qquad\times(2L+1)\,(-)^L\begin{pmatrix} l & l & L\\ m_1 & m_2 & -M\end{pmatrix}\begin{pmatrix} l & l & L\\ m_2 & m_4 & -M\end{pmatrix}
\end{aligned}$$

where the last expression is derived from the properties of the $3j$- and $6j$-symbols. The quantum numbers (nl) defining the subshell have been suppressed in the expression for the electron interaction integral on the left hand side of Equation (8.58).

At this point it is interesting to define electron pair annihilation and creation operators, $\pi\,(LM_LSM_S)$ and $\pi^\dagger(LM_LSM_S)$, such that $\pi^\dagger$ creates a normalized two-electron state with specified orbital and spin angular momentum, when applied to the vacuum state or a state where no electrons occupy the spin orbitals of the subshell. These operators are given by

$$\pi^\dagger(LM_LSM_S)=(-)^{M_L+M_S}\,[\tfrac{1}{2}(2L+1)\,(2S+1)]^{\frac{1}{2}} \times\sum\begin{pmatrix} l & l & L \\ m' & m & -M_L\end{pmatrix}\begin{pmatrix} \frac{1}{2} & \frac{1}{2} & S \\ \nu' & \nu & -M_S\end{pmatrix} a^\dagger_{m'\nu'}a^\dagger_{m\nu} \tag{8.59}$$

and by the adjoint equation.

The triplet operators do not exist for even L while the singlet operators are nonvanishing only for even L, as can be seen from the symmetry properties of the $3j$- symbols and the anticommutation relations of the field operators.

The hamiltonian operator in the reduced form for the subshell (nl) is then obtained as

$$H=\left[\epsilon(nl)-q(nl)F^0+\tfrac{1}{2}q(nl)\,\textstyle\sum_k F^k\begin{pmatrix} l & k & l \\ 0 & 0 & 0\end{pmatrix}^2\right]N_{nl}+\tfrac{1}{2}F^0N_{nl}(N_{nl}-1) +\textstyle\sum_{LM_LSM_S}V(L)\pi^\dagger(LM_LSM_S)\,\pi(LM_LSM_S), \tag{8.60}$$

where the elements $V(L)$ are the electronic repulsion energies in the pair states of angular momentum quantum number L:

$$V(L)=\textstyle\sum_{k\neq 0}F^k\,(2l+1)^2\begin{pmatrix} l & k & l \\ 0 & 0 & 0\end{pmatrix}^2\begin{Bmatrix} l & l & k \\ l & l & L\end{Bmatrix}(-)^L. \tag{8.61}$$

The average energy of the electrons in this single subshell is when computed with the density operator of Equation (7.26),

$$\langle H\rangle=\epsilon(nl)q(nl)-\tfrac{1}{2}q(nl)^2\left[F^0-\tfrac{1}{2}\textstyle\sum_k F^k\begin{pmatrix} l & k & l \\ 0 & 0 & 0\end{pmatrix}^2\right]. \tag{8.62}$$

The introduction of the pair operators $\pi^\dagger$ and π allows a simple way of demonstrating the fact that corresponding particle and hole configurations have the same terms and that similar energy expressions hold. This particle-hole symmetry has been well-known since the early work of Heisenberg. We will use a unitary transformation, which is generated by the pair operators $\pi^\dagger(0000)$ and $\pi(0000)$. The quantum numbers will be omitted in the following when we consider the totally symmetric ones.

A unitary operator U can be written in terms of a hermitian operator S as

$$U=\exp(iS) \tag{8.63}$$

and we study a particular choice,

$$S=\gamma\pi+\gamma^*\pi^\dagger, \tag{8.64}$$

where γ is an arbitrary complex number. The following commutator is needed,

$$[a_{m\nu}, S]=(-)^{l-m+\frac{1}{2}-\nu}\gamma^*(2l+1)^{-\frac{1}{2}}\, a^\dagger_{-m-\nu}. \tag{8.65}$$

The transformation of an operator A is given as

$$\tilde{A}=UAU^\dagger=A-i[A, S]+(i^2/2!)\,[\,[A, S], S]-(i^3/3!)\,[\,[\,[A, S], S], S]+\ldots \tag{8.66}$$

and we find that

$$\tilde{a}_{m\nu}=a_{m\nu}\cos|\gamma(2l+1)^{-\frac{1}{2}}|-(-)^{l-m+\frac{1}{2}-\nu}a^\dagger_{-m-\nu}\,(i\gamma^*/|\gamma|)\sin|\gamma(2l+1)^{-\frac{1}{2}}|. \tag{8.67}$$

A complete particle-hole transformation is effected when the absolute value of γ is $\frac{1}{2}\pi(2l+1)^{\frac{1}{2}}$ radians. The generators of the transformation commute with the total orbital and spin angular momenta and the transformation will connect only states with the same quantum numbers (LM_LSM_S).

The operators of the hamiltonian are transformed as follows under a complete particle-hole transformation

$$\tilde{N}_{nl}=4l+2-N_{nl}, \tag{8.68}$$

$$\tilde{\pi}^\dagger(LM_LSM_S)=(-)^{M_L+M_S}(i|\gamma|/\gamma)^2\,\pi\,(L-M_L\,S-M_S), \tag{8.69}$$

and we can conclude that

$$\begin{aligned}&\Sigma_{M_LM_S}\,\tilde{\pi}^\dagger(LM_LSM_S)\,\tilde{\pi}\,(LM_LSM_S)\\&=\Sigma_{M_LM_S}\,\pi^\dagger(LM_LSM_S)\,\pi(LM_LSM_S)+\Sigma_{M_LM_S}[\pi(LM_LSM_S),\,\pi^\dagger(LM_LSM_S)],\end{aligned} \tag{8.70}$$

where the last sum is equal to

$$(2L+1)\,(2S+1)\,[1-N_{nl}/(2l+1)],$$

thus it holds that apart from a term involving only the number operator the hamiltonian is invariant under the complete particle-hole transformation and that the relative energies of the terms of an $(nl)^q$-configuration are the same as those of the $(nl)^{4l+2-q}$-configuration.

One can generally suppose that the energy parameter $V(0)$ is the largest among the $V(L)$'s and we investigate the consequences of neglecting all but this one. Then it is clear that if there are states $|N, g\rangle$ which are characterized by the eigenvalue N of the number operator N_{nl}, and other quantum numbers g, and satisfy the equation

$$\pi|N, g\rangle=0 \tag{8.71}$$

it follows that the energy of such a state is dependent only upon N. Such a state is given the seniority quantum number N, and will be written $|N, N, g\rangle$.

Another set of states will satisfy the equation

$$(\pi)^{f+1}|\, N, N-2f, g\rangle=0, \tag{8.72}$$

and are given the seniority number $N-2f$, where it is implied that

$$(\pi)^{f}|\, N, N-2f, g\rangle\neq 0. \tag{8.73}$$

We will let g label the different eigenstates of the hamiltonian with common number of particles and seniority and define states

$$|\, N+2f, N, g\rangle=C_{fN}(\pi^{\dagger})^{f}|\, N, N, g\rangle, \tag{8.74}$$

which also are eigenstates of the hamiltonian. In Equation (8.74) C_{fN} denotes the normalization constant of the state,

$$\begin{aligned}|C_{fN}|^{-2}&=\langle N, N, g\,|\pi^{f}(\pi^{\dagger})^{f}|\, N, N, g\rangle\\ &=\langle N, N, g\,|\pi^{f-1}\,[\pi, (\pi^{\dagger})^{f}]\,|\, N, N, g\rangle\\ &=[f(2l+2-N-f)/(2l+1)]\,\langle N, N, g\,|\,\pi^{f-1}(\pi^{\dagger})^{f-1}|\, N, N, g\rangle\\ &=(f!)^{2}\binom{2l+1-N}{f}(2l+1)^{-f}.\end{aligned} \tag{8.75}$$

The calculation of the electronic interaction energy in the state is equally simple and gives

$$\begin{aligned}&\langle N+2f, N, g\,|\, V(0)\pi^{\dagger}\,\pi|\, N+2f, N, g\rangle\\ &\qquad=|C_{fN}|^{2}\,\langle N, N, g\,|\,\pi^{f}V(0)\,\pi^{\dagger}\pi\,(\pi^{\dagger})^{f}|\, N, N, g\rangle\\ &\qquad=V(0)f(2l+2-N-f)/(2l+1).\end{aligned} \tag{8.76}$$

It should be realized that the range of the parameters N and f can be restricted, due to particle-hole symmetry, so that

$$N+2f\leq 2l+1, \tag{8.77}$$

which implies that the largest interaction energy occurs when the difference between the number of electrons and the seniority number is the greatest. Racah introduced the seniority concept and found the seniority number to be an almost good quantum number. This is borne out in the previous analysis since the neglected terms in the hamiltonian are small. The pair creator $\pi^{\dagger}$ commutes with the angular momentum and spin operators for the electrons and will not change the corresponding quantum numbers for a state. A general particle-hole transformation will have matrix elements only between states of equal seniority, orbital and spin angular momentum.

Other two-particle operators can also be expressed in terms of the general pair creation and annihilation operators. We find that the total orbital angular momentum is

$$\bar{L}^2 = l(l+1)N_{nl} + 2l(l+1)(2l+1)\sum(-)^L \begin{Bmatrix} l & l & 1 \\ l & l & L \end{Bmatrix} \pi^\dagger(LM_LSM_S)\,\pi\,(LM_LSM_S) \tag{8.78}$$

and that the total spin is

$$\bar{S}^2 = (3/4)N_{nl} + 3\sum(-)^{S+1} \begin{Bmatrix} \frac{1}{2} & \frac{1}{2} & 1 \\ \frac{1}{2} & \frac{1}{2} & S \end{Bmatrix} \pi^\dagger\,(LM_LSM_S)\,\pi\,(LM_LSM_S). \tag{8.79}$$

The number of pairs of electrons is simply given as

$$\tfrac{1}{2}N_{nl}\,(N_{nl}-1) = \sum \pi^\dagger\,(LM_LSM_S)\,\pi\,(LM_LSM_S). \tag{8.80}$$

In effect it is possible to express any two-particle interactions in terms of these operators and the form will be as in Equation (8.60) with a more general notion given to the $V(L)$ parameters than what is implied in Equation (8.61). Such ideas have been used by Trees although expressed in other terms.

Electronegativity and Differential Ionization Potential

The use of ensemble average energy functionals admits considerations of the variation of the energy as functions of the occupation numbers $q(nl)$. We have already seen that the orbital energy $\epsilon(nl)$ equals the partial derivative $\partial\langle H\rangle/\partial q(nl)$ in Equation (8.50). A further analysis in terms of derivatives meets with difficulties since the orbital energies for the valence shell will be strongly varying with the charge in the shell. This is reflected in a plot of $\epsilon(4s)$ and $\epsilon(3d)$ for iron as in Fig. 3.

A noticeable feature of these curves is that they are closely approximated by a functional expression of the form

$$\epsilon(q) = (\alpha + \beta q)^{3/2}. \tag{8.81}$$

The quality of such a representation can be seen in Table 3. The whole range of possible occupations in the $3d$-shell of iron, cobalt, and nickel can be covered at an accuracy better than 6% with the formula (8.81).

It is a consequence of Equation (8.81) that the derivative

$$\partial\epsilon(nl)/\partial q(n'l') = \partial^2\langle H\rangle/\partial q(nl)\partial q(n'l'), \tag{8.82}$$

which would have an interpretation as an effective interaction parameter

$$\partial^2\langle H\rangle/\partial q(nl)\partial q(n'l') = V_{\text{eff}}(nl, n'l'), \tag{8.83}$$

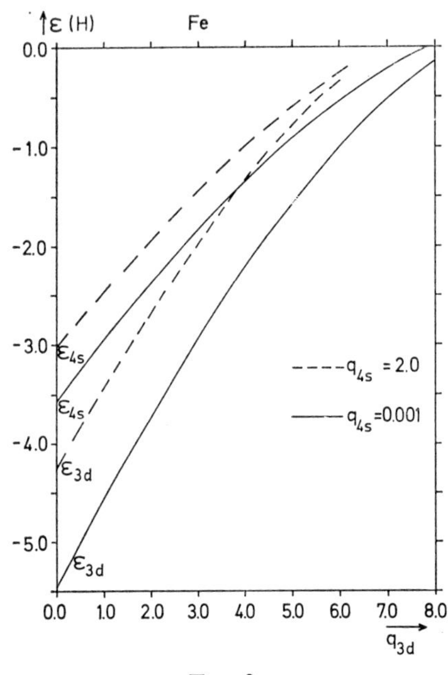

FIG. 3.

has a dependence on the occupation q of the form

$$V_{\text{eff}} \sim (\gamma + \delta q)^{\frac{1}{3}}. \tag{8.84}$$

Table 3 shows that the interaction integrals $V(3d, 3d)$ for iron, cobalt, and nickel can be closely approximated with a formula like (8.84) but with parameters that are relatively unrelated to those appropriate for orbital energies. The relation between the effective interaction parameters, defined in Equation (8.83) from the analogy to Landau's Fermi liquid theory, and the calculated integrals $V(nl, n'l')$ gives an idea of the screening mechanisms in atomic subshells. A more detailed study of these problems remains to be carried out.

Orbital energies in normal Hartree–Fock theory for closed shell systems have the interpretation as the negative of ionization potentials corresponding to poles of the Fourier transform of the electron propagator as discussed before. Open shell systems treated with the spherically symmetric ensemble averaging lead to orbital energies that are weighted means of ionization energies and electron affinities as will be further elaborated in Chapter 9. At this point we want to direct the attention to the concept of differential ionization energy as defined by Jørgensen. When a set of central field orbitals have been determined we can proceed as before by first order perturbation

TABLE 3. Analytical Approximations to Orbital Energy Parameters $\epsilon(3d)$ and Interaction Integrals $V(3d, 3d)$ for Iron, Cobalt, and Nickel Compared to Calculated Values. Energies in Hartrees

Fe $3d^q$ $\epsilon(3d) = -\frac{1}{2}[4\cdot9535 - 0\cdot5603q]^{3/2}$
$V(3d, 3d) = [1\cdot0349 - 0\cdot08209q]^{\frac{1}{2}}$

q	Formula $-\epsilon$	Formula V	Calculated $-\epsilon$	Calculated V
1	4·604	0·976	4·585	0·971
2	3·752	0·933	3·744	0·930
3	2·961	0·888	2·962	0·887
4	2·233	0·841	2·241	0·842
5	1·578	0·790	1·588	0·793
6	1·004	0·736	1·010	0·740
7	0·524	0·678	0·524	0·680
8	0·162	0·615	0·157	0·610

Fe $3d^q4s^2$ $\epsilon(3d) = -\frac{1}{2}[4\cdot1985 - 0\cdot5673q]^{3/2}$
$V(3d, 3d) = [0\cdot9846 - 0\cdot07801q]^{\frac{1}{2}}$

q	Formula $-\epsilon$	Formula V	Calculated $-\epsilon$	Calculated V
1	3·460	0·952	3·445	0·951
2	2·682	0·910	2·682	0·910
3	1·972	0·866	1·982	0·867
4	1·340	0·820	1·351	0·821
5	0·795	0·771	0·800	0·772
6	0·354	0·719	0·347	0·718

Co $3d^q$ $\epsilon(3d) = -\frac{1}{2}[5\cdot7165 - 0\cdot5715q]^{3/2}$
$V(3d, 3d) = [1\cdot2148 - 0\cdot08415q]^{\frac{1}{2}}$

q	Formula $-\epsilon$	Formula V	Calculated $-\epsilon$	Calculated V
1	5·835	1·063	5·787	1·059
2	4·890	1·023	4·861	1·020
3	4·003	0·981	3·991	0·979
4	3·177	0·937	3·179	0·937
5	2·417	0·891	2·428	0·893
6	1·730	0·843	1·744	0·846
7	1·124	0·791	1·135	0·795
8	0·612	0·736	0·614	0·737
9	0·217	0·676	0·208	0·671

TABLE 3. (*Cont.*).

Ni $3d^q$ $\quad \epsilon(3d) = -\frac{1}{2}[6\cdot4831 - 0\cdot5803q]^{3/2}$
$V(3d, 3d) = [1\cdot4046 - 0\cdot08637q]^{\frac{1}{2}}$

	Formula		Calculated	
q	$-\epsilon$	V	$-\epsilon$	V
1	7·171	1·148	7·106	1·145
2	6·140	1·110	6·096	1·107
3	5·163	1·070	5·141	1·068
4	4·245	1·029	4·240	1·028
5	3·389	0·986	3·398	0·987
6	2·600	0·941	2·618	0·943
7	1·883	0·894	1·903	0·897
8	1·249	0·845	1·263	0·848
9	0·708	0·792	0·709	0·793
10	0·280	0·735	0·266	0·730

theory to calculate term energies and particularly the average of a configuration. We obtain that for a set of integer occupations $q(nl)$ Slater's average energy for a configuration with N electrons is

$$E(N) = \langle H \rangle - \Sigma_{nl}\, q(nl)\, [4l+2-q(nl)]\, V(nl, nl)/(8l+2). \qquad (8.85)$$

It should be noticed that only open shells contribute to the sum on the right hand side. We denote the atomic number Z and consider an ion with charge x and define for *integer* x the differential ionization energy $I(y)$ by means of the relation

$$E(Z-x) = E(Z) + \int_0^x I(y)dy. \qquad (8.86)$$

It follows from Equation (8.85) that if we consider only one open valence subshell (nl) with $q(nl) = k$ corresponding to the neutral atom then from Equation (8.50)

$$\int_0^x I(y)dy = -\int_{k-x}^k \epsilon(nl)\, dq(nl) - \Delta(k, x) \qquad (8.87)$$

where Δ corresponds to the second term on the right hand side of Equation (8.85). The restriction to integer x can be removed in Equation (8.87) and derivatives can be taken to obtain that

$$I(x) = -\epsilon(nl)|q_{(nl)=k-y}x - \partial\Delta/\partial x. \qquad (8.88)$$

The term $-\partial\Delta/\partial x$ in this equation destroys the direct correspondence between the empirically defined differential ionization energy and the negative of the orbital energy $\epsilon(nl)$. It is relatively small but not without significance for comparisons with parmeterized functions $I(x)$.

Jørgensen has used a polynomial expansion for $I(x)$ like

$$I(x)=a_0+a_1x+a_2x^2+\dots \tag{8.89}$$

and has determined the three first coefficients from experimental ionization energies. In view of the functional forms for $\epsilon(nl)$ and $V(nl, nl)$ we cannot expect to obtain a convergent expansion like (8.89) except for very small x since the branch point of the square root occurs at negative x-values with small magnitude.

In order to make an attempt at comparing theoretically and experimentally derived parameters we have calculated for iron, cobalt and nickel the Taylor series expansion in the variable $\delta q=q(3d)-5$ of the right hand side of Equation (8.88) to second order. The expansion was then transformed to the form (8.89) in the variable x. The result of these calculations is given in Table 4

TABLE 4. Comparison between Empirical and Theoretical Differential Ionization Energy Parameters, $I(x)=a_0+a_1x+a_2x^2$, in Units of 1000 cm^{-1} for Iron, Cobalt, and Nickel

	Fe		Co		Ni	
	emp.	theor.	emp.	theor.	emp.	theor.
a_0	−46	−39	−36	−57	−28	−77
a_1	102	91	103	105	103	118
a_2	9	11	10	10	11	9

together with parameters from experiment. The agreement must be considered satisfactory since the Taylor series changes rapidly with the point around which the expansion is performed.

The Mulliken scale for electronegativity is based on the average of the ionization energy and the electron affinity and comes out to be

$$X=\frac{1}{2}\int_{-1}^{1} I(x)dx. \tag{8.90}$$

We see that for infinite systems like the electron gas it equals the negative of the chemical potential μ. For an open shell atom it would then naturally correspond to the orbital energy of Equation (8.50) for the partially filled subshell. This follows also from Equations (8.88) and (8.90) if we disregard the second term in Equation (8.88) and evaluate the integral (8.90) approximately.

The correlation between the valence shell orbital energy parameter $\epsilon(nl)$ and Mulliken's electronegativity is demonstrated in Fig. 4 for a range of atoms. This relation indicates that the theoretical parameter, $\epsilon(nl)$, might serve as an alternative measure of the electronegativity of an atom.

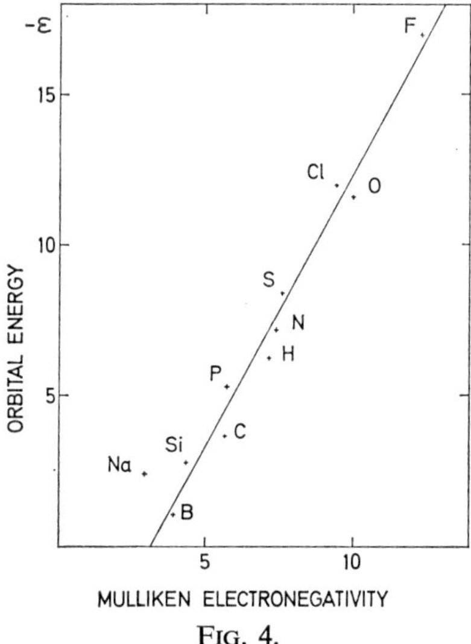

FIG. 4.

Problem 1

Express $\langle H \rangle$ of Equation (8.46) for the F atom in terms of $I(nl)$, F^k and G^k integrals.

Problem 2

Demonstrate that the width of the number operator $N_{\text{op}} = \sum_s n_s$ is

$$[\langle (N_{\text{op}} - \langle N_{\text{op}} \rangle)^2 \rangle]^{\frac{1}{2}} = [\textstyle\sum_s \langle n_s \rangle (1 - \langle n_s \rangle)]^{\frac{1}{2}},$$

when we average with the density operator

$$\rho = \textstyle\sum_s [1 - \langle n_s \rangle + (2 \langle n_s \rangle - 1) n_s].$$

Problem 3

(*a*) Introduce the complete set of projectors P_N corresponding to eigenvalues N of the number operator. Verify that the realization,

$$P_N = \int_0^1 dt \exp [2\pi\, it(N_{\text{op}} - N)],$$

of the projectors satisfy the relations

$$(N_{\text{op}} - N) P_N = 0,\ P_N^2 = P_N,\ \textstyle\sum_N P_N = 1.$$

(*b*) Introduce the density operator

$$\rho_N = P_N \rho / [Tr\ P_N \rho],$$

where ρ is the density operator in Problem 2. Use the notation

$$\langle \ldots \rangle_N = Tr(\rho_N \ldots) \text{ and } \langle n_{nlm_\nu} \rangle_N (4l+2) = q(nl),$$
$$Q(nl\, n'l') = \langle n_{nlm_\nu} n_{n'l'm'_{\nu'}} \rangle_N (4l+2)(4l'+2)$$

to show that

$$\langle H\rangle_N = \sum_{nl} I(nl)q(nl) + \tfrac{1}{2}\sum_{nln'l'} Q(nln'l')V(nl, n'l').$$

(*c*) Show that for a single open subshell (nl) the following relation holds

$$Q(nln'l') = q(nl)q(n'l') - \delta_{nn'}\delta_{ll'}\, q(nl)\,[4l+2-q(nl)]/(4l+1).$$

Notes and Bibliography

The reader, who is not familiar with 3*j*- and 6*j*- symbols can easily find out about their basic properties and fundamental relations from a great many different sources, e.g. E. P. Wigner, "Group Theory and its Application to the Quantum Mechanics of Atomic Spectra" (Academic Press, New York, 1959) or B. R. Judd, "Second Quantization and Atomic Spectroscopy" (The Johns Hopkins Press, Baltimore, 1967), where the relations between the reduced field operator matrix elements and the coefficients of fractional parentage can be found. The basic definitions of spherical harmonics as well as a wealth of information about the atomic central field problem can be found in E. U. Condon and G. H. Shortley, "The Theory of Atomic Spectra" (Cambridge University Press, 1953).

The atomic Hartree–Fock problem is described in considerable detail in J. C. Slater, "Quantum Theory of Atomic Structure" (McGraw-Hill Book Co., Vol. 1, Vol. 2, 1960). The theory of complex spectra also receives an in-depth treatment in the references mentioned above, but can maybe even better be studied in the original treatment by Slater (1929) in *Phys. Rev.* **34**, 1293, and by G. Racah (1942) in *Phys. Rev.* **61**, 186, and **62**, 438. The discussion of the particle-hole symmetry dates back to the early work of W. Heisenberg (1931) published in *Ann. Physik*, **10**, 888. Transformations of field operators generated by the totally symmetric pair operators has also been discussed by H. Watanabe (1966) in "Operator Methods in Ligand Field Theory" (Prentice-Hall, Inc., Englewood Cliffs), but in a different form from what has been done in this chapter. Racah in *Phys. Rev.* **63**, 367 (1943), introduced the seniority concept and R. E. Trees (1951) in *Phys. Rev.* **83**, 756 (1952), *ibid.*, **85**, 382, and *J. Opt. Soc. Amer.* **54**, 651 (1964), has developed ideas like those presented on pages 60–67. The work by C. K. Jørgensen on the idea of differential ionization potentials is presented in his book "Orbitals in Atoms and Molecules" (Academic Press, London, 1962).

Hartree–Fock calculations for the whole periodic system have been performed by J. B. Mann and are available in Reports Nos. LA-3690/1 from Los Alamos Scientific Laboratory, University of California, New Mexico, U.S.A.

APPENDIX I

Approximate Calculation of Atomic Transition Amplitudes

In this appendix we give a few examples of the approximate determination of matrix elements of the electron field operator. This is done with the purpose of showing the kind of approximations which are involved with the use of simple atomic orbitals in the representations of electron propagators. Atomic units are used.

1. *The Hydrogen Atom*

The ground state is degenerate and the corresponding manifold of states are

$$| 1s, 0, 0, \tfrac{1}{2}, \tfrac{1}{2}\rangle = \overset{\dagger}{a}_{100\frac{1}{2}}| \text{vac}\rangle, \tag{I.1}$$

$$| 1s, 0, 0, \tfrac{1}{2}, -\tfrac{1}{2}\rangle = \overset{\dagger}{a}_{100-\frac{1}{2}}| \text{vac}\rangle. \tag{I.2}$$

The amplitudes of interest are

$$\langle 1s, 0, 0, \tfrac{1}{2}, -\tfrac{1}{2} \,|\psi^\dagger(\xi)|\, \text{vac}\rangle = \overset{*}{u}_{100-\frac{1}{2}}(\xi) = u_{1s\beta}(\xi) \tag{I.3}$$

as defined in Equation (8.2) with the radial part, R_{10}, being the exact ground state solution of the radial Schrödinger equation for the hydrogen atom, and

$$\langle (1s)^2, 0, 0, 0, 0 \,|\psi^\dagger(\xi)|\, 1s, 0, 0, \tfrac{1}{2}, -\tfrac{1}{2}\rangle$$
$$= \int u_{1s\beta}(\xi') \langle (1s)^2, 0, 0, 0, 0 \,|\psi^\dagger(\xi)\, \psi^\dagger(\xi')|\, \text{vac}\rangle d\xi', \tag{I.4}$$

which we calculate with a simple wave function for H^-. Following Equation (4.33) we express the wave function as

$$\langle \text{vac} \,|\psi(\xi')\psi(\xi)|\, (1s)^2, 0, 0, 0, 0\rangle$$
$$= N \exp[-\mu(r+r')] \cosh[\nu(r-r')] \{\alpha(\zeta')\beta(\zeta) - \beta(\zeta')\alpha(\zeta)\}. \tag{I.5}$$

Shull and Löwdin, *J. Chem. Phys.* **25**, 1035 (1956), have variationally determined the parameters to be $\mu = 0{\cdot}6612$ and $\nu = 0{\cdot}3780$. It follows directly that

$$\langle (1s)^2, 0, 0, 0, 0 \,|\psi^\dagger(\xi)|\, 1s, 0, 0, \tfrac{1}{2}, -\tfrac{1}{2}\rangle$$
$$= -4N\pi\alpha^*(\zeta) \exp(-\mu r) [(1+\mu+\nu)^{-3} \exp(\nu r) + (1+\mu-\nu)^{-3} \exp(-\nu r)]. \tag{I.6}$$

This amplitude has the normalization integral

$$\int |\langle (1s)^2, 0, 0, 0, 0 \,|\psi^\dagger(\xi)|\, 1s, 0, 0, \tfrac{1}{2}, -\tfrac{1}{2}\rangle|^2 d\xi = 0{\cdot}695 \tag{I.7}$$

and overlaps the hydrogenic $1s$-orbital to almost 67% which in turn implies that

$$|\langle (1s)^2, 0, 0, 0, 0| \overset{\dagger}{a}_{100\frac{1}{2}} |1\, s, 0, 0, \tfrac{1}{2}, -\tfrac{1}{2}\rangle|^2 = 0{\cdot}445 \tag{I.8}$$

or 64% the value in Equation (I.7).

It is notable that the two largest terms in the sumrule

$$\sum_s \int u^*_{100\frac{1}{2}}(\xi) A(\xi, \xi'; s) u_{100\frac{1}{2}}(\xi') d\xi d\xi' = 1 \tag{I.9}$$

equals 72% of the total value.

A closer look at the amplitude in Equation (I.6) shows that the component, which dominates at large distances from the nucleus has an exponent, which is small compared to that of the other component. This reflects the fact that the probability amplitude for finding an electron far from the nucleus in H^-, and another in the $1s$ state, has an asymptotic functional form corresponding to the wave function of an electron with a small binding energy.

The numerical value of the asymptotic exponent as obtained by Shull and Löwdin is $\mu-\nu=0{\cdot}2832$, which can be compared to the theoretical value

$$(-2m\epsilon_s)^{\frac{1}{2}}=0{\cdot}2355.$$

2. *The Carbon and Nitrogen Atoms*

We introduce a quasi vacuum consisting of the singlet state for the core which is held fixed throughout the calculation. This state is denoted

$$|qv\rangle=|\,(1s)^2(2s)^2, 0, 0, 0, 0\rangle. \tag{I.10}$$

We limit the discussion to the lowest multiplets of configurations $(2p)$, $(2p)^2$, and $(2p)^3$.

States of interest can be calculated from approximate wave functions, such as

$$|\,2p, 1, 1, \tfrac{1}{2}, \tfrac{1}{2}\rangle=\int d\xi u(\xi)\psi^\dagger(\xi)|\,qv\rangle, \tag{I.11}$$

where $u(\xi)=\alpha(\zeta)Y_{11}(\theta, \phi)C(\mu)r\exp(-\mu r)$ with $C(\mu)$ being a normalization constant. Similarly we have

$$|(2p)^2, 1, 1, 1, 1\rangle=\int d\xi_1 d\xi_2 u(\xi_1, \xi_2)\,\psi^\dagger(\xi_1)\psi^\dagger(\xi_2)|\,qv\rangle, \tag{I.12}$$

where

$$u(\xi_1, \xi_2)=\alpha(\zeta_1)\alpha(\zeta_2)S(1, 2)C(\mu_1, \mu_2)\sum_P Pr_1r_2\exp(-\mu_1r_1-\mu_2r_2),$$

with $S(1, 2)$ being the antisymmetric angular part

$$S(1, 2)=\sum_{m_1, m_2} 3^{\frac{1}{2}}\begin{pmatrix}1 & 1 & 1\\ m_1 & m_2 & -1\end{pmatrix} Y_{1m_1}(1)Y_{1m_2}(2),$$

$C(\mu_1, \mu_2)$ the normalization constant, and $\sum_P P$ the symmetrizer. For the quartet state of $(2p)^3$ we obtain

$$|\,(2p)^3, 0, 0, \tfrac{3}{2}, \tfrac{3}{2}\,\rangle=\int d\xi_1 d\xi_2 d\xi_3 u(\xi_1, \xi_2, \xi_3)\,\psi^\dagger(\xi_1)\psi^\dagger(\xi_2)\psi^\dagger(\xi_3)\,|\,qv\rangle, \tag{I.13}$$

where

$$u(\xi_1, \xi_2, \xi_3)=\alpha(\zeta_1)\alpha(\zeta_2)\alpha(\zeta_3)S(1, 2, 3)\,C(\mu_1, \mu_2, \mu_3)\sum_P Pr_1\,r_2\,r_3 \times\exp(-\mu_1\,r_1-\mu_2\,r_2-\mu_3\,r_3)$$

with $S(1, 2, 3)$ being the antisymmetric angular part

$$S(1, 2, 3)=\sum_{m_1, m_2, m_3,}\begin{pmatrix}1 & 1 & 1\\ m_1 & m_2 & m_3\end{pmatrix} Y_{1m_1}(1)Y_{1m_2}(2)Y_{1m_3}(3),$$

and $C(\mu_1, \mu_2, \mu_3)$ the normalization constant. It should be observed that the spin and space parts are symmetric in these wave functions.

We define the normalized modified radial amplitudes

$$\overset{*}{P}_N(r) = r\,\langle \gamma NLS \| R_l\,(r) \| \gamma' N' L' S' \rangle\,[N(2L+1)\,(2S+1)]^{-\frac{1}{2}}. \qquad (\text{I}.14)$$

The exponents have been determined by variational calculations using the wave functions u for the separate ions. The following radial amplitudes obtain for carbon

$$P_5(r) = r^2\,5{\cdot}00546 \exp(-1{\cdot}798r) \qquad (\text{I}.15)$$

$$P_6(r) = r^2\,[3{\cdot}18457 \exp(-2{\cdot}016r) + 0{\cdot}82624 \exp(-1{\cdot}114r)] \qquad (\text{I}.16)$$

$$P_7(r) = r^2\,[0{\cdot}1549 \exp(-0{\cdot}628r) + 0{\cdot}5393 \exp(-1{\cdot}191r) + 2{\cdot}0578 \exp(-2{\cdot}061r)]. \qquad (\text{I}.17)$$

We give in Tables I.1 and I.2 the calculated exponents, total energies in atomic units or Hartrees, ionization potentials in electron volts.

TABLE I.1

Carbon:	N	4	5	6	7
	μ_1	0	1·798	1·114	0·628
	μ_2	0	0	2·016	1·191
	μ_3	0	0	0	2·061
(a.u.)	E_{tot}	−36·33054	−37·26425	−37·67795	−37·70538
(eV)	IP(calc)	—	25·41	11·26	0·75
(eV)	IP(exp)	—	24·38	11·26	1·25

TABLE I.2

Nitrogen:	N	4	5	6	7
	μ_1	0	2·345	2·644	2·787
	μ_2	0	0	1·624	1·732
	μ_3	0	0	0	1·220
(a.u.)	E_{tot}	−50·90343	−52·73425	−53·86133	−54·39581
(eV)	IP(calc.)	—	49·82	30·67	14·54
(eV)	IP(exp)	—	47·43	29·61	14·54

These results have been obtained with the assistance of Poul W. Thulstrup which we gratefully acknowledge.

CHAPTER 9

Atomic and Molecular Orbitals

The discussion in this chapter will concentrate on the field-theoretical interpretation and transcription of some approximate methods in the theory of molecular electronic structure. It is limited to the situation with a fixed nuclear framework and concerns the description of states that are close in energy to the normal state of the system. We adopt the view that the main features of the electronic structure of such states can be developed in terms of atomic orbital representations of operators and that only valence shell orbitals are to be considered. These assumptions are common to the method of molecular orbitals as linear combinations of atomic orbitals (MO-LCAO) and the valence bond method. The particular case of the Pariser–Parr–Pople model for conjugated systems is treated separately in Chapter 10.

Nonorthogonal Basis Set

The treatment of overlap in conjunction with the use of nonorthogonal basis orbitals deserves particular attention in a field-theoretical analysis. The definition of creation and annihilation operators and the calculation of the appropriate anticommutation rules is of basic importance for this development. A set of atomic orbitals, $[u_s(\xi)]$, is used to define creation operators

$$a_s^\dagger = \int u_s(\xi)\psi^\dagger(\xi)\, d\xi \tag{9.1}$$

and annihilation operators

$$a_s = \int u_s^*(\xi)\psi(\xi)\, d\xi. \tag{9.2}$$

It follows from Equation (4.2) that the anticommutation rules for these operators are

$$\begin{aligned} &[a_s, a_r]_+ = 0, \\ &[a_s^\dagger, a_r^\dagger]_+ = 0, \\ &[a_s, a_r^\dagger]_+ = \int u_s^*(\xi)u_r(\xi)\, d\xi \equiv \delta_{sr} + S_{sr}. \end{aligned} \tag{9.3}$$

An orthonormal basis set will have all matrix elements S_{sr} equal to zero.

The basis set defines a subspace of the total Hilbert space and it is convenient at this point to introduce the projection operator defined through the kernel,

$$D(\xi|\xi') = \Sigma_s u_s(\xi)\, u_s^*(\xi') + \Sigma_{sr} u_r(\xi) T_{sr} u_r^*(\xi'), \tag{9.4}$$

where the elements T_{sr} are related to the overlaps S_{sr} through the system

$$T_{sr} + S_{sr} + \Sigma_t\, S_{st} T_{tr} = 0. \tag{9.5}$$

The relation (9.5) can alternatively be expressed such that the matrix **T** is the deviation of the inverse of the matrix $(\mathbf{1}+\mathbf{S})$ from a unit matrix:

$$\mathbf{T} = (\mathbf{1}+\mathbf{S})^{-1} - \mathbf{1}. \tag{9.6}$$

The projection operator defines the components of the field operators that will be of concern to us in the treatment,

$$\psi(\xi) \to \psi_D(\xi) = \int D(\xi|\xi')\psi(\xi')\, d\xi' = \Sigma_s u_s(\xi)\, a_s + \Sigma_{sr} u_s(\xi) T_{sr} a_r, \tag{9.7}$$

and it is the purpose of the present development to examine the limitations this poses on the general theory.

The truncated field operator of Equation (9.7) is an example of an approximate representation of an operator. Correspondingly there will be approximate representations of the states of interest to us. It will be assumed that such states can be generated from a common reference state, analogous to the vacuum state of Equation (4.12). The reference state should have the property that

$$\psi_D(\xi)|\, \mathrm{ref}\rangle = 0, \tag{9.8}$$

and need not be specified further. A general state in the space to be considered can be expressed as

$$|\Phi\rangle = \int d(12\ldots N)\psi_D^\dagger(1)\psi_D^\dagger(2)\ldots\psi_D^\dagger(N)|\,\mathrm{ref}\rangle\, \Phi(12\ldots N), \tag{9.9}$$

or as

$$|\Phi\rangle = \Sigma\, a_r^\dagger a_s^\dagger \ldots a_t^\dagger\, |\,\mathrm{ref}\rangle\, \Phi_{rs\ldots t}, \tag{9.10}$$

where $\Phi(12\ldots N)$ and $\Phi_{rs\ldots t}$ are appropriate amplitudes corresponding to the wave function in a continuous and discreet representation respectively. The dimension of this space is 2^M where M is the number of basis functions in the set $[u_s(\xi)]$.

The problem can now be formulated as one of finding the appropriate mapping from the states of physical interest to the model states given in Equations (9.9) and (9.10). The transform will obviously be dependent upon the reference state and the basis set.

Green's Function Considerations

Relevant observations can be made on the appropriateness of a limited basis set and a subsequent choice of the hamiltonian from the formal structure of the electron propagator for a free atom. We wish to examine the Green's function $G(\xi, \xi'; \epsilon)$ of Equation (8.30) in order to extract information on an optimal choice of valence orbitals and related energy parameters. The limitation of the basis set leads us to consider the outer projection of the propagator

$$G_D(\xi, \xi'; \epsilon) = \langle\langle \psi_D(\xi); \psi_D^\dagger(\xi') \rangle\rangle_\epsilon = \int d(12) D(\xi|1) G(1, 2; \epsilon) D(2|\xi'), \quad (9.11)$$

and its corresponding spectral densities $A_D(\xi, \xi'; \epsilon)$. The projected spectral densities obey the sum rule

$$\int d\epsilon\, A_D(\xi, \xi'; \epsilon) = D(\xi|\xi'), \quad (9.12)$$

as compared to Equation (8.38).

We choose to treat explicitly the case of an atom, the low-lying states of which can be described as arising from orbital configurations of the form $[s^{q(s)} p^{q(p)}]$. The basis set will then have 8 elements: $(\nu = \pm\frac{1}{2})$

$$\begin{aligned} u_{s\nu}(\xi) &= R_0(r) Y_{00}(\theta, \phi)\, \delta_{\nu\zeta}, \\ u_{x\nu}(\xi) &= R_1(r)\, [Y_{1-1}(\theta, \phi) - Y_{11}(\theta, \phi)]\, \delta_{\nu\zeta}/\sqrt{2}, \\ u_{y\nu}(\xi) &= R_1(r)\, [Y_{1-1}(\theta, \phi) + Y_{11}(\theta, \phi)]\, i\delta_{\nu\zeta}/\sqrt{2}, \\ u_{z\nu}(\xi) &= R_1(r)\, Y_{10}(\theta, \phi)\, \delta_{\nu\zeta}. \end{aligned} \quad (9.13)$$

The projection operator kernel $D(\xi|\xi')$ is then of the form

$$D(\xi|\xi') = \delta_{\zeta\zeta'}\, [R_0(r)\, \overset{*}{R_0}(r') + 3P_1(\cos\Theta)\, R_1(r)\, \overset{*}{R_1}(r')]/4\pi, \quad (9.14)$$

where Θ is defined as in Equation (8.16). It follows from Equation (9.11) that the spectral density will appear as

$$A_D(\xi, \xi'; \epsilon) = \delta_{\zeta\zeta'}\, [R_0(r) A_0(\epsilon)\, \overset{*}{R_0}(r') + 3P_1(\cos\Theta)\, R_1(r) A_1(\epsilon) \overset{*}{R_1}(r')]/4\pi, \quad (9.15)$$

where the "orbital" spectral densities $A_0(\epsilon)$ and $A_1(\epsilon)$ both sum to unity.

We have seen in Chapter 7 that the use of moment expansions can be helpful in the determination of propagators and thus for the spectral densities. Such expansions are obtained from expectation values of repeated commutators, see Equation (7.17), and we introduce the following notation:

$$\begin{aligned} a_r^{(k)} &= [a_r^{(k-1)}, H], \\ a_r^{(0)} &= a_r. \end{aligned} \quad (9.16)$$

Thus we obtain that the spectral densities $A_0(\epsilon)$ and $A_1(\epsilon)$ satisfy the following sum rules,

$$\int d\epsilon\epsilon^k A_0(\epsilon) = \langle [a_{s\frac{1}{2}}^{(k)}, a_{s\frac{1}{2}}^{\dagger}]_+ \rangle \equiv W_0(k), \tag{9.17}$$

$$\int d\epsilon\epsilon^k A_1(\epsilon) = \langle [a_{x\frac{1}{2}}^{(k)}, a_{x\frac{1}{2}}^{\dagger}]_+ \rangle \equiv W_1(k). \tag{9.18}$$

The expectation values are supposed to be taken over a rotationally invariant ensemble as in Chapter 8. The moments $W_l(k)$ will clearly depend upon the choice of ensemble and upon the overlap between the radial amplitudes of Equation (8.14) and the basis functions $R_l(r)$. When it holds, as is often the case, that these overlaps are large only for a few states one can expect to be able to calculate approximate values for the moments from experimental energy levels by means of fractional parentage and vector coupling coefficients. We might similarly assume that moments are known from atomic calculations. The moments will now be used to establish a model hamiltonian for the atomic valence shell.

A model hamiltonian will have a structure similar to the full hamiltonian as given by Equations (4.42), (4.4), and (4.38), but it could possibly include terms of higher order in products of annihilation and creation operators. We will consider only such operators which can be said to contain a one-particle part and an electron interaction term. The number of independent matrix elements will be considerably reduced by symmetry considerations and by imposing further conditions on compatibility with other operator representatives. It is clear that the form of the spectral density requires that the hamiltonian commutes with the orbital angular momentum and spin operators. These are given in the limited basis as

$$\bar{L} = i\bar{e}_x \sum_\nu (a_{z\nu}^{\dagger} a_{y\nu} - a_{y\nu}^{\dagger} a_{z\nu}) + i\bar{e}_y \sum_\nu (a_{x\nu}^{\dagger} a_{z\nu} - a_{z\nu}^{\dagger} a_{x\nu}) + i\bar{e}_z \sum_\nu (a_{y\nu}^{\dagger} a_{x\nu} - a_{x\nu}^{\dagger} a_{y\nu}), \tag{9.19}$$

and

$$\bar{S} = \bar{S}_s + \bar{S}_x + \bar{S}_y + \bar{S}_z, \tag{9.20}$$

where the definition (4.50) has been used. The only one-particle part that commutes with $\bar{L}$ and $\bar{S}$ is a linear combination of the number operators N_l (cf. Equation (8.21)):

$$H = \alpha_0 N_0 + \alpha_1 N_1 = \alpha_0 \sum_\nu a_{s\nu}^{\dagger} a_{s\nu} + \alpha_1 \sum_\nu (a_{x\nu}^{\dagger} a_{x\nu} + a_{y\nu}^{\dagger} a_{y\nu} + a_{z\nu}^{\dagger} a_{z\nu}). \tag{9.21}$$

Supposing we were to limit our model hamiltonian to this expression we would then choose α_l equal to $W_l(1)$, which would guarantee that the first moment of the spectral density is in agreement with the experimentally determined value. This corresponds to the assumption that α_l equals the orbital electronegativity parameter with reversed sign.

The electron interaction term is more complicated than the one-particle part. A convenient means for its examination are the pair creation operators listed in Table 5 and their adjoints. Their definition derives from Equation (8.59) but is here given in a cartesian form.

TABLE 5. Pair Creation Operators for a Set of s- and p-Operators in Cartesian Form

γ	L	M_L	S	M_S	$\pi^\dagger(\gamma L M_L S M_S)$
s^2	0	0	0	0	$a^\dagger_{s\frac{1}{2}} a^\dagger_{s-\frac{1}{2}}$
sp	1	x	0	0	$(a^\dagger_{s\frac{1}{2}} a^\dagger_{x-\frac{1}{2}} - a^\dagger_{s-\frac{1}{2}} a^\dagger_{x\frac{1}{2}})/\sqrt{2}$
sp	1	y	0	0	$(a^\dagger_{s\frac{1}{2}} a^\dagger_{y-\frac{1}{2}} - a^\dagger_{s-\frac{1}{2}} a^\dagger_{y\frac{1}{2}})/\sqrt{2}$
sp	1	z	0	0	$(a^\dagger_{s\frac{1}{2}} a^\dagger_{z-\frac{1}{2}} - a^\dagger_{s-\frac{1}{2}} a^\dagger_{z\frac{1}{2}})/\sqrt{2}$
p^2	0	0	0	0	$(a^\dagger_{x\frac{1}{2}} a^\dagger_{x-\frac{1}{2}} + a^\dagger_{y\frac{1}{2}} a^\dagger_{y-\frac{1}{2}} + a^\dagger_{z\frac{1}{2}} a^\dagger_{z-\frac{1}{2}})/\sqrt{3}$
p^2	2	xy	0	0	$(a^\dagger_{x\frac{1}{2}} a^\dagger_{y-\frac{1}{2}} - a^\dagger_{x-\frac{1}{2}} a^\dagger_{y\frac{1}{2}})/\sqrt{2}$
p^2	2	yz	0	0	$(a^\dagger_{y\frac{1}{2}} a^\dagger_{z-\frac{1}{2}} - a^\dagger_{y-\frac{1}{2}} a^\dagger_{z\frac{1}{2}})/\sqrt{2}$
p^2	2	zx	0	0	$(a^\dagger_{z\frac{1}{2}} a^\dagger_{x-\frac{1}{2}} - a^\dagger_{z-\frac{1}{2}} a^\dagger_{x\frac{1}{2}})/\sqrt{2}$
p^2	2	x^2-y^2	0	0	$(a^\dagger_{x\frac{1}{2}} a^\dagger_{x-\frac{1}{2}} - a^\dagger_{y\frac{1}{2}} a^\dagger_{y-\frac{1}{2}})/\sqrt{2}$
p^2	2	z^2	0	0	$(2a^\dagger_{z\frac{1}{2}} a^\dagger_{z-\frac{1}{2}} - a^\dagger_{x\frac{1}{2}} a^\dagger_{x-\frac{1}{2}} - a^\dagger_{y\frac{1}{2}} a^\dagger_{y-\frac{1}{2}})/\sqrt{6}$
sp	1	x	1	1	$a^\dagger_{s\frac{1}{2}} a^\dagger_{x\frac{1}{2}}$
sp	1	y	1	1	$a^\dagger_{s\frac{1}{2}} a^\dagger_{y\frac{1}{2}}$
sp	1	z	1	1	$a^\dagger_{s\frac{1}{2}} a^\dagger_{z\frac{1}{2}}$
p^2	1	x	1	1	$a^\dagger_{y\frac{1}{2}} a^\dagger_{z\frac{1}{2}}$
p^2	1	y	1	1	$a^\dagger_{z\frac{1}{2}} a^\dagger_{x\frac{1}{2}}$
p^2	1	z	1	1	$a^\dagger_{x\frac{1}{2}} a^\dagger_{y\frac{1}{2}}$

Note: Other operators for $S=1$ can easily be obtained from the ones that are given from the relation

$$[S_-, \pi^\dagger(SM_S)] = \pi^\dagger(SM_S-1)\,[(S+M_S)(S-M_S+1)]^{\frac{1}{2}}$$

for tensor operators.

Corresponding to the pair operators there will be an interaction parameter $V(\gamma LS)$ in the interaction hamiltonian as in Equation (8.60). There is also a possibility to introduce parameters that couple pair states of equal L and S but different γ's. We require that the two-particle part should commute with the *parity* operator,

$$P = \exp\,(i\pi\,\textstyle\sum_l lN_l), \tag{9.22}$$

and this leaves us only one possible extra parameter to be called $V(s^2, p^2)$. The interaction hamiltonian will then have the form

$$H_{\text{int}} = \sum V(\gamma LS)\,\pi^\dagger(\gamma LM_LSM_S)\,\pi(\gamma LM_LSM_S) + V(s^2, p^2)\,[\pi^\dagger(s^2 0000)\,\pi(p^2 0000) + \pi^\dagger(p^2 0000)\pi(s^2 0000)]. \tag{9.23}$$

TABLE 6. Electron Interaction Parameters for Pair States of s- and p-Orbitals from Slater–Condon Integrals

γ	L	S	$V(\gamma LS)$
s^2	0	0	$F^0(s,s)$
sp	1	0	$F^0(s,p)+(1/3)G^1(s,p)$
sp	1	1	$F^0(s,p)-(1/3)G^1(s,p)$
p^2	1	1	$F^0(p,p)-(1/5)F^2(p,p)$
p^2	2	0	$F^0(p,p)+(1/25)F^2(p,p)$
p^2	0	0	$F^0(p,p)+(2/5)F^2(p,p)$
	s^2, p^2		$G^1(s,p)/\sqrt{3}$

The parameters are listed in Table 6 for the case that they should be calculated from Slater–Condon F- and G-integrals.

We notice that there are only five independent Slater–Condon integrals while the most general expression contains seven free parameters.

Accurate atomic energy levels could be determined from a hamiltonian such as $H+H_{\text{int}}$ from Equations (9.21) and (9.23) with nine parameters at our disposal. In order to describe transitions between these levels it is necessary to consider a representative for the electric dipole moment operator in the basis. We will use the form

$$\bar{R}=e\int d\xi\psi_D^{\dagger}(\xi)\,\bar{r}\psi_D(\xi) \tag{9.24}$$

and the notation

$$\mu=(e/\sqrt{3})\int drr^3R_0(r)R_1(r), \tag{9.25}$$

to obtain

$$R_x=\mu\,\textstyle\sum_\nu\,(a_{s\nu}^{\dagger}a_{x\nu}+a_{x\nu}^{\dagger}a_{s\nu}), \tag{9.26}$$

for one component of the dipole moment operator. We notice immediately that R_x, R_y and R_z do not commute with one another, a typical result from the truncation of the basis. The dipole velocity operator for a system would normally be derived from the commutator of $\bar{R}$ and the hamiltonian and we find that

$$\begin{aligned}\dot{R}_x&=-i[R_x,H+H_{\text{int}}]\\&=-i\mu(\alpha_1-\alpha_0)\textstyle\sum_\nu\,(a_{s\nu}^{\dagger}a_{x\nu}-a_{x\nu}^{\dagger}a_{s\nu})-i[R_x,H_{\text{int}}].\end{aligned} \tag{9.27}$$

The second term on the right hand side of Equation (9.21) will be a two-particle contribution to the dipole velocity operator. We impose the restriction that the dipole velocity operator should be represented as a one-particle operator as in the case of a complete basis. Thus we require that

$$[R_x,H_{\text{int}}]=0. \tag{9.28}$$

It turns out that only two independent interaction parameters $V(\gamma LS)$ remain when Equation (9.28) is satisfied.

A convenient form can be found for H_{int} when the condition (9.28) holds. It is clear that the operators $N = N_0 + N_1$ and $\bar{S}$ will commute with $\bar{R}$ and that we can express the interaction in terms of the operators N^2 and $\bar{S}^2$. The model hamiltonian with these constraints has the form

$$H_{\text{atom}} = \sum_l \alpha_l N_l + \tfrac{1}{2}\gamma N(N-1) - J(\bar{S}^2 - \tfrac{3}{4}N), \qquad (9.29)$$

where γ represents an average Coulomb-type integral while J appears as an exchange integral. The calculation of moments from the operator (9.29) gives the expression

$$W_l(1) = \alpha_l + \gamma \langle N \rangle - (\gamma - \tfrac{3}{2}J) \langle N_l \rangle / (4l+2), \qquad (9.30)$$

which now are explicitly dependent upon the ensemble used for averaging.

The second moments are more involved but can be evaluated straightforwardly. Thus we find that

$$W_l(2) = W_l^2(1) + (\gamma - \tfrac{3}{2}J)^2 \langle N_l \rangle (4l+2-\langle N_l \rangle)(4l+2)^{-2} + (2\gamma - J)J \langle \bar{S}.\bar{S}_l \rangle /(2l+1) + J^2 \langle \bar{S}^2 \rangle, \qquad (9.31)$$

where

$$\bar{S}_0 = \bar{S}_s, \qquad (9.32)$$

and

$$\bar{S}_1 = (\bar{S}_x + \bar{S}_y + \bar{S}_z)/3. \qquad (9.33)$$

Since there are four parameters in the model hamiltonian, they can be determined from the four moments $W_0(1)$, $W_1(1)$, $W_0(2)$, and $W_1(2)$. The detailed calculation of the moments depends on the number of available energy levels and how one wishes to weigh them through an ensemble average. According to Equation (8.18) we obtain the moments for a term as

$$W_l(\gamma LS, k) = g \sum_{\gamma' L' S'} [E(\gamma' L' S') - E(\gamma LS)]^k | \langle \gamma LS \| a_l^\dagger \| \gamma' L' S' \rangle |^2 + g \sum_{\gamma' L' S'} [E(\gamma LS) - E(\gamma' L' S')]^k | \langle \gamma' L' S' \| a_l^\dagger \| \gamma LS \rangle |^2, \qquad (9.34)$$

where the weight factor g is given as

$$g^{-1} = (4l+2)(2L+1)(2S+1). \qquad (9.35)$$

The appropriate weights are listed in Table 7 for the configurations under consideration here with the reduced matrix elements of creation operators obtained from the simple orbital picture.

TABLE 7. Weights for the Calculation of Moments of the Spectral Densities $A_l(\epsilon)$ for s^2p^n Configurations

		$\frac{|\langle p^nLS\|a_1^\dagger\|p^{n-1}L'S'\rangle|^2}{6(2L+1)(2S+1)}$			$\frac{|\langle p^{n+1}L'S'\|a_1^\dagger\|p^nLS\rangle|^2}{6(2L+1)(2S+1)}$		
n	$LL'=$	0	1	2	0	1	2
0	0	0	0	0	0	1	0
1	1	1/6	0	0	1/18	1/2	5/18
2	0	0	1/3	0	0	2/3	0
2	1	0	1/3	0	2/9	1/6	5/18
2	2	0	1/3	0	0	1/6	1/2
3	0	0	1/2	0	0	1/2	0
3	1	1/9	5/36	1/4	1/9	5/36	1/4
3	2	0	1/4	1/4	0	1/4	1/4
4	0	0	2/3	0	0	1/3	0
4	1	2/9	1/6	5/18	0	1/3	0
4	2	0	1/6	1/2	0	1/3	0
5	1	1/18	1/2	5/18	1/6	0	0
6	0	0	1	0	0	0	0

$$\frac{|\langle s^2p^nLS\|a_s^\dagger\|sp^nLS'\rangle|^2}{2(2L+1)(2S+1)}=\frac{2S'+1}{4S+2}$$

Note that for s^2p^n configurations there is a unique spin multiplicity for each L-value.

The calculation of hamiltonian parameters from moments of the spectral density for the electron propagator offers a more systematic approach to the general problem of assigning weights in a fitting procedure for energy levels. We notice that, for the simple hamiltonian obtained when J is neglected in Equation (9.29), we recover the Coulomb parameter γ as the difference between ionization energy and electron affinity of the atom.

A Simple Model Hamiltonian

The analysis of Green's functions permits us to establish model hamiltonians with the property of giving approximately correct propagators when used in the equation of motion. We want to demonstrate a particularly simple model in order to get familiar with some concepts in molecular orbital theory as phrased in the Green's function language. The Hartree–Fock approximation will be used and we wish to determine the Fock operator matrix elements,

$$f_{rs}=\langle[\,[a_r,\,H_{\text{tot}}],\,a_s^\dagger]_+\rangle. \tag{9.36}$$

A guide to this calculation can be obtained from a consideration of the limit of separated atoms,

$$f_{rs} \xrightarrow[\text{separated atoms}]{} \delta_{rs} W_r(1), \tag{9.37}$$

which can be derived from an effective hamiltonian of the form

$$H_{\text{eff}} = \sum_r W_r(1)\, a_r^\dagger a_r, \tag{9.38}$$

containing only the first moments of the atomic spectral density functions. Let us assume that the model hamiltonian (9.38) applies also at interatomic distances in molecules. The Fock operator is then, when Equation (9.3) applies,

$$f_{rs} = \delta_{rs} W_r + S_{rs}(W_r + W_s) + \sum_t S_{rt} W_t S_{ts}, \tag{9.39}$$

where the specific notation for the first moment has been dropped. This result is closely similar to the widely used approximations suggested from different arguments by Mulliken and by Wolfsberg and Helmholz. The Green's function equation of motion is according to Equation (5.8)

$$E G_{rs}(E) = \delta_{rs} + S_{rs} + W_r G_{rs}(E) + \sum_t S_{rt} W_t G_{ts}(E), \tag{9.40}$$

and we can directly write down the solution in matrix form as

$$\mathbf{G}(E) = [E(\mathbf{1}+\mathbf{T}) - \mathbf{W}]^{-1}, \tag{9.41}$$

where $\mathbf{T}$ is the matrix defined by Equation (9.6) and $\mathbf{W}$ is a diagonal matrix. It is often the case that the elements W_r all are negative and we might then write

$$W_r = -\chi_r^2, \tag{9.42}$$

where the elements χ_r form a positive definite diagonal matrix $\boldsymbol{\chi}$. An alternative form can then be given for the electron propagator,

$$\mathbf{G}(E) = \boldsymbol{\chi}^{-1}\, [E\mathbf{1} + \boldsymbol{\chi}\,(\mathbf{1}+\mathbf{S})\boldsymbol{\chi}]^{-1} \boldsymbol{\chi}\,(\mathbf{1}+\mathbf{S}), \tag{9.43}$$

where it is seen that the poles of the Green's function, the molecular orbital energies, are obtained as the eigenvalues of the negative definite matrix with elements $-\chi_r(\delta_{rs}+S_{rs})\chi_s$. The molecular orbital coefficients can be inferred from the residues at the poles and the spectral representation.

The very simple assumptions behind the model hamiltonian (9.38) lead to acceptable values for electron binding energies for molecular species as can be seen from the examples in Table 8 where a comparison is made with more involved molecular orbital calculations and with results from photoelectron spectroscopy.

TABLE 8. Electron Binding Energies in Molecules SF_6(O_h Symmetry) and CF_4 (T_d Symmetry) as determined by Calculations Compared with Observed Values from Photoelectron Spectroscopy. All Energies in eV

Molecule	Symmetry	(a)	(b)	(c)	(d)	Obs
SF_6	e_g	19·4	17·5	17·8	18·1	~16
	t_{1u}	19·0	16·8	21·5	18·5	
	t_{1g}	18·2	15·9	22·4	17·9	17·3
	t_{2u}	19·4	16·8	23·1	19·1	18·7
	t_{2g}	22·2	18·8	24·2	21·7	19·9
	t_{1u}	24·7	21·8	29·0	23·5	22·9
	a_{1g}	29·6	26·7	30·8	27·5	27·0
	e_g	45·4	35·6	43·1	41·4	39·3
	t_{1u}	46·9	36·5	50·8	47·4	41·2
	a_{1g}	50·4	39·3	57·0	57·4	44·2
CF_4	t_2		16·2	20·5	18·2	16·1
	t_1		15·3	22·5	18·3	17·4
	e		16·8	23·6	20·5	18·5
	t_2		20·8	28·5	23·5	22·2
	a_1		23·4	29·8	25·9	25·1
	t_2		35·0	48·5	45·7	40·3
	a_1		37·7	55·0	56·3	43·8

(a) Hartree–Fock calculation with gaussian basis set. U. Gelius, B. Roos, and P. Siegbahn, *Chem. Phys. Letters* **4**, 471 (1970) as quoted in (b).
(b) "Muffin-tin" potential, one-electron self-consistent calculation. J. W. D. Connolly and K. H. Johnson, *Chem. Phys. Letters* **10**, 616 (1971), for SF_6, J. W. D. Connolly, *Int. 3. Quantum Chem.* **S6**, 201 (1972), for CF_4.
(c) CNDO approximation: K. Siegbahn, C. Nordling, G. Johansson, J. Hedman, P. F. Hedén, K. Hamrin, U. Gelius, T. Bergmark, L. O. Werme, R. Manne, and Y. Baer, *ESCA Applied to Free Molecules* (North-Holland Publishing Co., Amsterdam, 1969). This reference contains also the observed values.
(d) Calculated from the poles of the electron propagator (9·43) with W_r's taken as free atom Hartree–Fock orbital energies. Overlap integrals were calculated from the corresponding atomic orbitals given by E. Clementi, *Tables of Atomic Functions* Supp. to *IBM Journal of Research and Development* **9**, (1965).

Calculation of Electronic Indices from Green's Functions

The electron propagator can be used to calculate the charge distribution and other electronic indices that are relevant for a discussion of ground state properties of molecules. We will examine in detail the Mulliken population analysis, which attempts to relate the charge distribution to a partial occupation of atomic orbitals and valence shells.

The number operator in the limited basis is obtained from Equation (9.7) as

$$N_{\text{op}} = \sum_r \left[a_r^\dagger + \sum_s a_s^\dagger T_{sr}\right] a_r, \tag{9.44}$$

and we define population operators as

$$N_{r,\,\mathrm{op}}=[a_r^{\dagger}+\textstyle\sum_s a_s^{\dagger}T_{sr}]\,a_r. \tag{9.45}$$

These operators commute among themselves, they are idempotent, but they are not hermitian.

The gross population of the spin orbital r is defined to be the expectation value of $N_{r,\,\mathrm{op}}$. According to Equation (5.22) we have that

$$N(r)=\langle N_{r,\,\mathrm{op}}\rangle=\frac{1}{2\pi i}\int_C dE\,\textstyle\sum_s G_{rs}(E)\,[\delta_{sr}+T_{sr}]. \tag{9.46}$$

We obtain for the particular model of the previous paragraph that

$$N(r)=(2\pi i)^{-1}\int_C dE\,[E\mathbf{1}+\boldsymbol{\chi}\,(\mathbf{1}+\mathbf{S})\boldsymbol{\chi}]_{rr}^{-1} \tag{9.47}$$

which is the result that would have been obtained if for an orthonormal set of orbitals we had chosen the Fock operator matrix elements as

$$-\chi_r\chi_s(\delta_{rs}+S_{rs}).$$

The gross atomic population which gives a measure of the formal charge on an atom in a molecule is defined to be the sum of all $N(r)$ where r is a spin orbital on the atom in question.

The population numbers $N(r)$ are clearly not physical observables in a strict sense but they serve as electronic indices. They may also be used to calculate an approximation to the dipole moment of the molecule in a point charge model where the dipole moment operator is approximated as

$$\bar{R}=\textstyle\sum_s eN_{s,\,\mathrm{op}}\int d\xi\,\bar{r}\mid u_s(\xi)|^2. \tag{9.48}$$

Other electronic indices are the bond orders which may be correlated with bond distances and bond energies. Their definition in the case of nonorthogonal orbitals has been subject to much discussion. The simple model which has been examined here offers a new definition in terms of the matrix used in Equation (9.47). We suggest use in the present case of bond orders

$$p_{rs}=(2\pi i)^{-1}\int_C dE[E\mathbf{1}+\boldsymbol{\chi}\,(\mathbf{1}+\mathbf{S})\boldsymbol{\chi}]_{rs}^{-1}. \tag{9.49}$$

This is in line with the definitions used in the Hückel theory but with a more extended basis set.

Orthogonalized Atomic Orbitals and Zero Differential Overlap Models

A variety of models for molecular electronic structure have been developed where the overlap problem is circumvented in that formally orthogonalized orbitals are employed. We examine in some detail in Chapter 10 the so-called

Pariser–Parr–Pople model for π-electron systems. Here we will put forth some of the more general features of such models. The symmetric orthogonalization procedure introduced by Löwdin offers a unique way of transforming a set of atomic orbitals to an orthonormal set where each member has as large an overlap with a corresponding member in the original set as is consistent with orthonormality.

The symmetrically orthogonalized set corresponding to $[u_s(\xi)]$ is denoted $[\tilde{u}_s(\xi)]$ and we have that

$$u_s(\xi)=\sum_t \tilde{u}_t(\xi)c_{ts}, \tag{9.50}$$

where

$$c_{ts}=\int d\xi\, \overset{*}{\tilde{u}_t}(\xi)u_s(\xi). \tag{9.51}$$

These relations lead to the equation

$$\delta_{rs}+S_{rs}=\sum_t \overset{*}{c}_{tr}c_{ts}. \tag{9.52}$$

A solution to this equation can be written in terms of the unitary matrix $[X_{sk}]$ which diagonalizes the overlap matrix,

$$\sum_{rs} X^{\dagger}_{kr}S_{rs}X_{sk'}=\lambda_k\delta_{kk'}, \tag{9.53}$$

and another unitary matrix $[Y_{rk}]$, as

$$c_{ts}=\sum_k Y_{tk}[1+\lambda_k]^{\frac{1}{2}}X^{\dagger}_{ks}. \tag{9.54}$$

The maximum overlap property between members of the two sets takes the form that the expression,

$$\Phi=\sum_s \int d\xi|u_s(\xi)-\tilde{u}_s(\xi)|^2=\sum_s [2-c_{ss}-\overset{*}{c}_{ss}], \tag{9.55}$$

should be a minimum when the unitary matrix $[Y_{tk}]$ is varied. It follows that the optimum choice is

$$Y_{tk}=X_{tk}, \tag{9.56}$$

in which case we obtain in matrix form that

$$\mathbf{c}=(\mathbf{1}+\mathbf{S})^{\frac{1}{2}}. \tag{9.57}$$

Acording to the definition (9.2) and the relation (9.50) we have the transformation for the field operators

$$a^{\dagger}_s=\sum_t \tilde{a}^{\dagger}_t\, c_{ts}. \tag{9.58}$$

The transformed operators $\tilde{a}_s$ and $\tilde{a}^{\dagger}_t$ satisfy anticommutation relations appropriate for an orthonormal basis set.

The orthonormal set $[\tilde{u}_s]$ is now employed for representations of the hamiltonian and other relevant operators. The orthogonality of basis functions that are formally associated with different atomic centers is taken, in approximate theories, as an indication that other integrals involving the product density of these spin orbitals are small and can be neglected to a first approximation. This means essentially that the charge density operator can be simplified as follows

$$q(\bar{r})=e\sum_{\text{spin}}\psi_D^\dagger(\xi)\psi_D(\xi)=e\sum_{\text{spin},\,s,\,t}\tilde{u}_s^*(\xi)\tilde{u}_t(\xi)\tilde{a}_s^\dagger\tilde{a}_t\simeq\sum_A q_A(\bar{r}), \qquad (9.59)$$

where $q_A(\bar{r})$ is an atomic charge density operator defined as the partial sum where orbital indices s and t both belong to the set of spin orbitals on atom A. Two center terms are neglected and the approximation has been termed Neglect of Diatomic Differential Overlap (NDDO).

A more drastic approximation than the NDDO is the Complete Neglect of Differential Overlap (CNDO), where all cross products in the sum in Equation (9.59) are omitted and the atomic charge density operator is obtained as

$$q_A(\bar{r})=e\sum_{\text{spin}}\sum_s^A|\tilde{u}_s(\xi)|^2\tilde{a}_s^\dagger\tilde{a}_s. \qquad (9.60)$$

Both the NDDO and the CNDO approximations apply only to matrix elements involving the charge density operator, other arguments have to be used for matrix elements of for instance the current density and the kinetic energy operators.

Before going on with our considerations we shall examine the density operator $q_A(\bar{r})$ somewhat more. It is clear that with an incomplete basis set it can no longer be assured that the charge density operator at point $\bar{r}$ commutes with the operator at $\bar{r}'\neq\bar{r}$. The CNDO approximation restores this feature of the full theory to the model theory. The cost for this property is rather high in that then also the dipole moment operator will have a representation where so-called atomic dipoles are neglected. That is to say that matrix elements of the position operator between orbitals on the same atom are omitted even though these are important for the determination of transition moments, for instance between s- and p-orbitals. The form of $q_A(\bar{r})$ should preferably be invariant under unitary transformations among the basis orbitals of atom A and this implies that the CNDO approximation must be used in a form such as

$$q_A(\bar{r})=e|\tilde{u}_A|^2\sum_s^A\tilde{a}_s^\dagger\tilde{a}_s=e|\tilde{u}_A|^2N_A, \qquad (9.61)$$

where N_A is the number operator for atom A, and where $|\tilde{u}_A|^2$ denotes a representative orbital density on atom A.

When the CNDO approximation to the charge density operator is adopted we obtain a representation for the electron interaction term of the hamiltonian as

$$H_{\text{int}}\cong\tfrac{1}{2}\sum\gamma_{AB}N_A(N_B-\delta_{AB}). \qquad (9.62)$$

The one-electron terms in the total hamiltonian cannot be expressed only in terms of the charge density and one adopts the general form

$$H \cong \sum h_{rs}\, \tilde{a}_r^\dagger \tilde{a}_s, \tag{9.63}$$

together with the prescription that off-diagonal elements h_{rs} are proportional to the overlap integrals S_{rs}. The remaining diagonal elements h_{ss} are most often chosen such that atomic properties are reproduced when overlaps S_{rs} and thus the co-called hopping terms h_{rs} in the hamiltonian are omitted.

The Fock matrix elements can be simply expressed when the total hamiltonian is taken as the sum $H+H_{\text{int}}$ from Equations (9.62) and (9.63):

$$f_{rs} = \langle [\,[\tilde{a}_r,\, H+H_{\text{int}}],\ \tilde{a}_s^\dagger]_+\rangle = h_{rs} - \gamma_{AB}\,\langle \tilde{a}_s^\dagger\, \tilde{a}_r\rangle + \delta_{sr} \sum_C \gamma_{AC}\,\langle N_C\rangle, \tag{9.64}$$

where spin orbitals r and s are associated with atoms A and B respectively.

Neglect of off-diagonal elements leads to that the number operators N_A will commute with the total hamiltonian and that bond orders $\langle \tilde{a}_s^\dagger \tilde{a}_r\rangle$ vanish. Then it follows that the expectation values $\langle N_C\rangle$ take integer values equal to the normal number of occupied valence orbitals in an isolated atom and we find that

$$\langle N_C\rangle \rightarrow z_C. \tag{9.65}$$

The identification with the "separated" atoms case leads then by comparison with Equation (9.30) to the formula for the diagonal Fock matrix elements,

$$f_{rr} = \alpha_r + \gamma_{AA}(z_A - \langle \tilde{a}_r^\dagger \tilde{a}_r\rangle) + \sum_C \gamma_{AC}(\langle N_C\rangle - z_C). \tag{9.66}$$

The discussion in this section has pointed out some properties of approximate molecular orbital calculations based on parametrized representations of the hamiltonian. Some operator relationships that are violated when an incomplete basis is used can be re-established at the expense of approximations in the evaluation of integrals. Atomic parameters are derived from considerations of a separated atom limit. Interatomic parameters are commonly associated with overlap integrals and other functions of the interatomic distance. For instance it is often assumed that when r is a spin orbital on atom A and s is one on B then

$$h_{rs} = \tfrac{1}{2}\,(\beta_A^0 + \beta_B^0) S_{rs}, \tag{9.67}$$

where the β^0-parameters are gained from comparisons with theoretical or experimental data for selected molecules. The form (9.60) for off-diagonal one-electron matrix elements leads us to the final formula for the hamiltonian in the CNDO approximation, with a different formulation we call it the CNDO model hamiltonian,

$$\begin{aligned} H(\text{CNDO}) = \text{constant} &+ \textstyle\sum_s \alpha_s \tilde{a}_s^\dagger \tilde{a}_s + \sum_{rs} \tfrac{1}{2}(\beta_A^0 + \beta_B^0)\, S_{rs} \tilde{a}_r^\dagger \tilde{a}_s \\ &+ \tfrac{1}{2} \textstyle\sum_{AB} \gamma_{AB}(N_A - z_A)\,(N_B - z_B) + \sum_A \gamma_{AA} N_A\,(z_A - \tfrac{1}{2}). \end{aligned} \tag{9.68}$$

The constant contribution in the hamiltonian is insignificant in the equations of motion but can serve to define an appropriate average value for some reference state. The next chapter contains a discussion of the Pariser–Parr–Pople model, which is similar to the CNDO-model in its formal structure, from a somewhat different point of view which illuminates some other aspects of model hamiltonians related to atomic orbital representations of field operators.

We saw in connection with the establishment of the atomic model hamiltonian that this could contain average exchange integrals and still commute with the dipole moment operator. The strict application of the CNDO approximation of Equation (9.60) gives a purely diagonal representation of the dipole moment but it has been the experience that intra-atomic dipole integrals give significant contributions to both transition moments and permanent dipole moments in molecules. This indicates that one would favor the NDDO approximation to CNDO. The argument leading to Equation (9.28) showed on the other hand that only electron interaction terms constructed from operators N_A and $\bar{S}_A$ could be considered. Thus we are lead to consider a model hamiltonian which contains intra- and inter-atomic exchange integrals J_{AB} such as

$$H(\text{CNDOX}) = H(\text{CNDO}) - \Sigma_{AB} J_{AB} \bar{S}_A . \bar{S}_B. \tag{9.69}$$

The interatomic exchange integrals would be zero in any approximation where diatomic differential overlap is neglected while the intra-atomic exchange are included in the approximation scheme that is called Intermediate Neglect of Differential Overlap (INDO). The form (9.69) is more restricted with regard to the number of electron interaction parameters than the NDDO approximation. These restrictions are imposed by Equation (9.28) as soon as the dipole moment operator cannot be expressed simply in terms of the N_A's.

Problem 1

Show that when a finite atomic orbital basis is used, for instance the set of s-, p_x-, p_y- and p_z-orbitals on an atom, and the operator representatives of the x- and y-components of the position operator are calculated, these operators do not commute.

Problem 2

Show that the operators in Problem 1 commute with the CNDO approximation to the charge density operator of Equation (9.61).

Notes and Bibliography

Many of the arguments and approximations in molecular orbital theory based on linear combinations of atomic orbitals are those of R. S. Mulliken. Particularly, we wish to refer here to his review in *J. chim. phys.* **46**, 497 and 675 from 1949. The population analysis is put forth in *J. Chem. Phys.* **23**, 1833 (1955). Orthogonalized atomic orbitals as they are defined in this chapter were first suggested by P. O. Löwdin in 1950 (*J. Chem. Phys.* **18**, 365). The NDDO, CNDO and INDO approximations are reviewed by J. A. Pople and D. L. Beveridge in "Approximate Molecular Orbital Theory" (McGraw-Hill, 1970).

CHAPTER 10

The Model Hamiltonian of Pariser, Parr and Pople

Propagator or Green's function methods are employed in this chapter for an analysis of the many-electron problem in planar unsaturated molecules as specified by the Pariser–Parr–Pople model. A derivation of the model in second quantization language and a discussion of its parameters in terms of many-electron theory demonstrates the nature of the approximations which are involved. Applications are presented for the case of weakly interacting atoms and it is shown that a decoupling procedure for Green's functions proposed by the authors is capable of giving a correct description of this case.

The propagator approach has the advantage of giving direct information on transition energies and amplitudes from a reference state but it suffers from a lack of sufficient knowledge of the analytical conditions that should be satisfied, a deficiency which is shared with the density matrix approach. A comparison with state vector descriptions is generally quite difficult beyond the Hartree–Fock approximation. In spite of this we feel that the propagator approach is worth investigating as an alternative to other current efforts. Particularly it appears that the semi-empirical approach to the electronic structure of molecules can be advantageously treated within a field-theoretical framework. It has been shown by Linderberg that such considerations can lead to useful relations between matrix elements in the Pariser–Parr–Pople model.

The molecular orbital method is a very flexible and often a successful tool in the analysis of the electronic structure dependent properties of molecules. Its drawbacks are intimately connected with the treatment of superposition of configurations, which has been pointed out by Slater. The conclusion is that the molecular orbital method is not satisfactory when the overlap between the relevant valence orbitals is smaller than 1/2 for adjacent atoms. This result was reached also by Coulson and Fischer in their well-known study of the hydrogen molecule, and it should be kept in mind when the method is applied to π-electron systems where the typical overlap integrals are in the range 1/3–1/4. Fischer–Hjalmars has recently presented evidence for the insufficiency of the Pariser–Parr–Pople model when the overlap exceeds

1/2 and it appears to us that the construction of approximate eigenstates to the Pariser–Parr–Pople hamiltonian from the molecular orbital method thus requires considerable justification. The instabilities of the normal ground states in the molecular orbital method that are noticed when the random phase approximation or the alternant molecular orbital method are applied to these problems give further support to this view.

An extension of the simplest picture in the molecular orbital method to include superposition of configurations has been attempted by several authors and Cizek has managed to give an effective formulation of the problem and carry out some applications. The drawback of such an approach is the need to perform separate calculations for different states and the difficulty of justification that the separate calculations are similar in accuracy. The Green's functions give direct information on energy differences of sets of states and the approximation procedures are generally not designed to improve the description of a particular state.

As an alternative to the method discussed above we have chosen to follow up some ideas presented by Hubbard for the study of narrow energy bands in solids with the view to examine magnetism. The main line in this work is to analyse in detail the structure of the many-electron problem for the case of separated atoms, that is the limit of zero band width.

It is shown that a useful separation of intra- and interatomic characteristics can be accomplished. A perturbation theory which begins from the atomic limit appears to be of a considerably more complex nature than what can profitably be used in calculations on actual systems. We have therefore attempted to develop the decoupling procedure for the truncation of the chains of equations obtained for double-time Green's functions as given by Zubarev. These efforts have led to a suggestion for a general technique to achieve decoupling in a systematic way.

We present in the next section a transcription into second quantization language the derivation of the π-electron approximation and the limitations in the implementation of it which was developed by Pariser and Parr and by Pople. The second section contains an analysis of the separated atom limit with special regard given to elementary excitation operators and to the derivation of an equivalent Heisenberg hamiltonian for the determination of the ground state energy. Interatomic terms are treated to full extent in the third part of the chapter by means of the decoupling procedure mentioned above and some consequences of the technique are discussed. The equations of motion which are derived require the calculation of various expectation values and they are given in the fourth section, where some problems in this connection are discussed as well. An application of the proposed scheme to a linear chain is given and comparisons with calculations according to the alternant molecular orbital method are presented in the fifth section.

Reduction to the Pariser–Parr–Pople Model

We expound in this section some steps which are necessary in order to arrive at the simplification of the many-electron hamiltonian for a molecule which has been called the Pariser–Parr–Pople model. It is assumed that the nuclear framework of the molecule is invariant under the operations of the group C_s which is the group of the identity and a reflection operation. This may not be true for all molecules to which the subsequent development is applied and the magnitude of the perturbation caused by the non-invariant part of the nuclear potential should then be examined. The electron field operator $\psi(\xi)$, defined at the space–spin point $\xi=(\bar{r}\zeta)$, will be expressed as the sum of two components, each of which transforms according to an irreducible representation of C_s:

$$\psi(\xi)=\psi(a', \xi)+\psi(a'', \xi) \tag{10.1}$$

The a'- and a''-components are the σ- and π-components respectively in the conventional nomenclature.

We repeat the anticommutation relations for electron field operators:

$$\begin{aligned}
&[\psi(\xi), \psi(\xi')]_+=\psi(\xi)\psi(\xi')+\psi(\xi')\psi(\xi)=0,\\
&[\psi(\xi), \psi^\dagger(\xi')]_+=\psi(\xi)\,\psi^\dagger(\xi')+\psi^\dagger(\xi')\psi(\xi)=\delta(\xi-\xi'),\\
&[\psi^\dagger(\xi), \psi^\dagger(\xi')]_+=\psi^\dagger(\xi)\psi^\dagger(\xi')+\psi^\dagger(\xi)\psi^\dagger(\xi)=0.
\end{aligned} \tag{10.2}$$

Two operators are particularly important for the description of the various states of concern, the number operator and the hamiltonian.

The former is

$$N_{\text{op}}=\int d\xi\psi^\dagger\,(\xi)\psi(\xi)=N(a')+N(a''), \tag{10.3}$$

where $N(a')$ is the operator for the number of electrons in the "σ-field" and $N(a'')$ has the similar interpretation for the "π-field".

The hamiltonian will be given as the sum of several terms:

$$\begin{aligned}
H&=H_0+H_1+H_2,\\
H_0&=\tfrac{1}{2}e^2\textstyle\sum' Z_g Z_h/|\bar{R}_g-\bar{R}_h|,\\
H_1&=\int d\xi\psi^\dagger(\xi)\,[-(1/2m)\nabla^2-e^2\textstyle\sum Z_g/|\bar{r}-\bar{R}_g|]\,\psi(\xi)\\
&=H_1(a')+H_1(a''),
\end{aligned} \tag{10.4}$$

$$\begin{aligned}
H_2&=\tfrac{1}{2}e^2\int d\xi\,d\xi'\,|\bar{r}-\bar{r}'|^{-1}\,\psi^\dagger(\xi)\psi^\dagger(\xi')\psi(\xi')\psi(\xi)\\
&=H_2(a', a', a', a')+2H_2(a'', a', a', a'')+2H_2(a'', a', a'', a')\\
&\quad+H_2(a', a', a'', a'')+H_2(a'', a'', a', a')+H_2(a'', a'', a'', a'').
\end{aligned} \tag{10.5}$$

Loosely speaking we have divided the energy operator into parts representing in order, the mutual repulsion of nuclei, the kinetic and potential energy of the σ- and π-"fields", the self-interaction of the σ-"field", the coulomb and exchange interaction between σ- and π-"fields", the energy of transfer of electrons from π to σ and reverse, and, finally, the self-interaction of the π-"field". There will be an extra term in H_1 when the group C_s is only approximately the group of the molecule. Units have been chosen such that $\hbar = 1$.

Electron transfer terms, the fourth and fifth part of H_2, do not commute with $N(a')$ and $N(a'')$ separately and prevent us from considering the number of "π-electrons" as a constant of the motion. We conclude that the omission of these terms is equivalent to the *ansatz* considered by Lykos and Parr. A perturbation theory argument indicates that the effects of the omission is small when the ratio of exchange-like integrals between a'- and a''-orbitals or reverse is small. It appears that this approximation is more valid than some approximations introduced in the following.

A second step considered by Lykos and Parr is the "freezing" of the σ-skeleton. In second quantization it corresponds to a neglect of the operator nature of $\psi(a', \xi)$ and the replacement of products like $\psi^\dagger(a', \xi')\psi(a', \xi)$ in the hamiltonian with their expectation values, assuming that such average values are not dependent on the particulars of the states under consideration. This is a crucial point in the establishment of the Pariser–Parr–Pople model and requires special attention. Perturbation theory shows this time that the approximation is reasonable if the polarizability of the σ-skeleton is small. An experimental determination of the polarizability of graphite layers by Taft and Philipp has convincingly demonstrated that for long wavelength optical excitations it is meaningful to consider the electron system as consisting of two parts, a σ- and a π-system. The analysis also demonstrates that the polarizability for the σ-skeleton is not small and that dispersive effects are not important for energies less than about 10–16 eV. The total effect of the σ-system is to shift the resonance energy of the π-system to lower energy, which is to say that the neglect of dynamical effects of the σ-skeleton can be formally compensated for by a reduction of the electronic interaction within the π-system. Such a reduction or screening can be looked upon as a dielectric effect from the σ-medium, but it is too much of a simplification to introduce a simple dielectric factor in the electron repulsion integrals.

Screening of electronic interactions can be qualitatively understood, but cannot easily be subject to numerical estimates. The difficulty arises from the inhomogeneous nature of the medium, since we want to describe the reduced interaction at distances comparable to bond distances in the molecule. It is neither sufficient to consider only local effects of screening nor to screen the various wavelengths in the Fourier transform of the Coulomb interaction independently. Hubbard formulated an integral equation for the screened interaction but only very approximate solutions have been attempted so far. We will demonstrate that some numerical evidence can be given for a determination of interaction integrals on the basis of a screening theory.

The neglect of dynamical effects of the σ-skeleton makes it possible to incorporate the second and third term of H_2 into $H_1(a'')$ to obtain $H_{\text{core}}(a'')$ in the usual nomenclature. The first term of H_2 combines with H_0 and $H_1(a')$ to a new constant, H_0'. Thus we are left with the effective hamiltonian for the π-electron system

$$H(a'') = H_0' + H_{\text{core}}(a'') + H_2(a'', a'', a'', a''). \tag{10.6}$$

The precise definition of $H_{\text{core}}(a'')$ is

$$\begin{aligned} H_{\text{core}}(a'') = &\int d\xi \psi^\dagger(a'', \xi) \left[-(1/2m)\nabla^2 + V_{\text{core}}(\bar{r})\right] \psi(a'', \xi) \\ &- e^2 \int d\xi\, d\xi' \psi^\dagger(a'', \xi) \langle \psi^\dagger(a', \xi')\psi(a', \xi)\rangle \psi(a'', \xi') \left|\bar{r} - \bar{r}'\right|^{-1}, \end{aligned} \tag{10.7}$$

$$V_{\text{core}}(\bar{r}) = -e^2 \sum Z_g/|\bar{r} - \bar{R}_g| + e^2 \int d\xi' \langle \psi^\dagger(a', \xi')\psi(a', \xi')\rangle \left|\bar{r} - \bar{r}'\right|^{-1}. \tag{10.8}$$

The core-potential is here taken as the Coulomb potential from the nuclei and the electron density in the a'-system, and the exchange potential is represented separately.

It is worth mentioning at this stage that Equation (10.6) is a complete definition of $H(a'')$ and it cannot without extra specifications be given a configuration space representation. The field operator $\psi(a'', \xi)$ contains the supplementary condition that in the configuration space picture of the states there will only be π-orbitals in the wave function.

One further step is required to reach the model we are looking for, the approximation of linear combination of atomic orbitals. The field operator will be expanded in a symmetrically orthonormalized set of atomic orbitals of a''-symmetry and we assume, as is usual, that each orbital is associated with a separate atomic center of the molecule:

$$\psi(a'', \xi) = \sum u_{r\nu}(\xi) a_{r\nu}, \tag{10.9}$$

$$a_{r\nu} = \int d\xi u_{r\nu}^*(\xi)\, \psi(a'', \xi). \tag{10.10}$$

The operators $a_{r\nu}$ are annihilation operators for electrons in the π-orbital on atom r with spin component ν. We invoke the zero differential overlap approximation, that is, the spin orbital product $u_{r\nu}(\xi)u_{s\nu'}(\xi)$ is put equal to zero for $r \neq s$. This leads to

$$\psi^\dagger(a'', \xi)\psi(a'', \xi) = \sum |u_{r\nu}(\xi)|^2 n_{r\nu}, \tag{10.11}$$

where $n_{r\nu} = a_{r\nu}^\dagger a_{r\nu}$ as before is the number operator for the spin orbital $(r\nu)$. The orbital expansions (10.9) and (10.11) are inserted in the hamiltonian

(10.6) and give the hamiltonian of the Pariser–Parr–Pople model:

$$H(\text{PPP}) = H_0' + \sum \alpha_r n_{r\nu} + \sum' \beta_{rs} a_{r\nu}^\dagger a_{s\nu} + \tfrac{1}{2} \sum' \gamma_{rs} n_{r\nu} n_{s\nu'}. \tag{10.12}$$

The primes on the summation signs indicate the omission of terms where the operators would correspond to the same spin orbital. Thus there is neither a term with β_{rr} nor with $n_{r\nu}n_{r\nu}$, but there is one with $n_{r\nu}n_{r-\nu}$. The parameters α_r and β_{rs} are matrix elements of H_{core} between spin orbitals $u_{r\nu}(\xi)$ and $u_{s\nu}(\xi)$, and γ_{rs} is the electron repulsion integral between spin orbital densities $|u_{r\nu'}(\xi)|^2$ and $|u_{s\nu}(\xi)|^2$.

We should observe that it is an immediate consequence of the form of the hamiltonian (10.12) that excitation energies for transitions between states of equal number of π-electrons involve only differences of the α's and γ's as has been proved by several authors. The operator given by

$$\begin{aligned} \delta H &= \delta\alpha N(a'') + \tfrac{1}{2}\delta\gamma N(a'')\,[N(a'') - 1] \\ &= \delta\alpha \sum n_{r\nu} + \tfrac{1}{2}\delta\gamma \sum' n_{r\nu} n_{s\nu'} \end{aligned} \tag{10.13}$$

can be added to H(PPP) with the effect of changing it to an equivalent form where all α's and γ's are increased uniformly with $\delta\alpha$ and $\delta\gamma$ respectively. Moreover, δH is a constant within each manifold of eigenstates belonging to a definite eigenvalue of $N(a'')$, thus it follows that energy eigenvalue differences within such a manifold are independent of $\delta\alpha$ and $\delta\gamma$.

The theorem discussed in the previous paragraph allows us to draw some conclusions with regard to the appropriate magnitude of the electron repulsion integrals. According to Hubbard's theory for screening of electronic interactions we expect that γ_{rs} represents the potential felt at atom r, averaged over the charge distribution $|u_r|^2$, from an electron with the charge distribution $|u_s|^2$ *and* its surrounding polarization field. We can assume, judging from the experimental results on graphite by Taft and Philipp that the polarization will be essentially instantaneous on the time-scale appropriate for the π-electrons. It will also have very little effect. This seems also to be the consensus of most empirical determinations of these parameters. A modification due to dielectric effects of the electron repulsion parameters influences only one-center integrals and nearest neighbor integrals and the changes will be noticeable in spectra through the differences of these integrals to second neighbor values. For benzene we notice that these differences as given by Simmons in various approximations are generally scaled down by a factor $\frac{1}{2}$, which would correspond to an effective value of the dielectric constant for the σ-medium equal to 2, which compares well with the optical value 2·5 found by Taft and Philipp. The technique afforded by Hubbard's theory of screening appears more satisfactory than simple perturbation theory arguments. One could suggest that the evaluation of the electron repulsion integrals be based on the assumption that the charge distributions should be determined from a more diffuse atomic π-orbital which is computed for a valence state configuration. There are certain drawbacks to this attempt since there will be an increase in the nearest neighbor integrals accompanying this decrease of localization and a greater difficulty in the justification of the zero differential overlap approximation.

Limit of Separated Atoms

We examine in some detail in this section the case when the off-diagonal matrix elements of the core operator, β_{rs}, are small in comparison to the electron repulsion integrals γ_{rr}. This corresponds to the physical situation of separated atoms but it should be realized that other matrix elements will also depend on the relative positions of the nuclei, both through the core-operator and through the symmetric orthonormalization procedure of the orbitals in the expansion (10.9). The constant term of H(PPP) will be omitted in the following and the third term will be considered as a perturbation while the second and fourth parts will constitute the unperturbed hamiltonian in the subsequent analysis.

The hamiltonian for the limit of separated atoms can be written as

$$\begin{aligned} H_{\text{atoms}} &= \textstyle\sum_r H_r + \frac{1}{2} \sum_{r \neq s} \gamma_{rs}(n_{r+} + n_{r-})(n_{s+} + n_{s-}), \\ H_r &= \alpha_r(n_{r+} + n_{r-}) + \gamma_{rr} n_{r+} n_{r-}. \end{aligned} \tag{10.14}$$

According to the anticommutation relations we have that

$$[a_{r\nu}, a^{\dagger}_{s\nu}]_+ = \delta_{\nu\nu'} \delta_{rs}, \tag{10.15}$$

while all other anticommutators vanish. Thus all operators $n_{r\nu}$ commute and we can conclude in particular that all operators N_r,

$$N_r = n_{r+} + n_{r-}, \tag{10.16}$$

commute with one another and with H_{atoms}.

The unperturbed energy eigenvalues are given in terms of the eigenvalues of N_r's, which are 0, 1, or 2. An atom can have none, one, or two electrons in the π-orbital. The states with one electron in the orbital are degenerate and a large number of degenerate states will correspond to each unperturbed value. Some of these degeneracies can be classified according to the theory of coupling of spin angular momentum.

Let us consider the operators

$$\begin{aligned} S_{rx} &= \tfrac{1}{2} [a^{\dagger}_{r+} a_{r-} + a^{\dagger}_{r-} a_{r+}], \\ S_{ry} &= \tfrac{1}{2} i [a^{\dagger}_{r+} a_{r-} - a^{\dagger}_{r-} a_{r+}], \\ S_{rz} &= \tfrac{1}{2} [a^{\dagger}_{r+} a_{r+} - a^{\dagger}_{r-} a_{r-}]. \end{aligned} \tag{10.17}$$

They satisfy the commutation relations for the components of angular momentum,

$$[S_{rx}, S_{ry}] = iS_{rz}, \qquad x, y, z, \text{ cyclic} \tag{10.18}$$

and we find that the total spin is given by

$$\bar{S}_r^2 = S_{rx}^2 + S_{ry}^2 + S_{rz}^2 = \tfrac{3}{4}N_r(2-N_r). \tag{10.19}$$

Within each manifold of states characterized by eigenvalues of the N_r's there will be a unique correspondence between the operators $\bar{S}_r$ and ordinary spin operators and they will be interpreted as the spin operators for atoms. The different spin components for an atom are all constants of the motion in the unperturbed case. Spin coupling will be important when the perturbation is introduced.

Before going into detail on spin coupling mechanisms, we will study the Heisenberg equation of motion for the electron annihilation operators in an unperturbed case. Hubbard found that the two components $n_{r-\nu}a_{r\nu}$ and $(1-n_{r-\nu})a_{r\nu}$ of the operator $a_{r\nu}$ were particularly simple to analyse since they develop harmonically in time when the hamiltonian is H_r:

$$i\frac{d}{dt}(1-n_{r-\nu})\, a_{r\nu} = [\,(1-n_{r-\nu})\, a_{r\nu}, H_r] = \alpha_r(1-n_{r-\nu})\, a_{r\nu}, \tag{10.20}$$

$$i\frac{d}{dt}n_{r-\nu}a_{r\nu} = [n_{r-\nu}a_{r\nu}, H_r] = (\alpha_r + \gamma_{rr})\, n_{r-\nu}a_{r\nu}. \tag{10.21}$$

The solution of Equations (10.20) and (10.21) give the result that

$$a_{r\nu}(t) = [1 + n_{r-\nu}(0)(\exp(-i\gamma_{rr}t) - 1)]\, a_{r\nu}(0) \exp(-i\alpha_r t). \tag{10.22}$$

Equation (10.22) can also be written in the form

$$a_{r\nu}(t) = a_{r\nu}(0) \exp[-it(\alpha_r + \gamma_{rr}n_{r-\nu})] \tag{10.23}$$

which is less explicit. Similarly we can solve the equation of motion when the total unperturbed hamiltonian is used:

$$\begin{aligned} a_{r\nu}(t) &= \exp(itH_{\text{atoms}})\, a_{r\nu}(0) \exp(-itH_{\text{atoms}}) \\ &= a_{r\nu}(0) \exp[-it(\alpha_r + \gamma_{rr}n_{r-\nu}) + \textstyle\sum_{s\neq r}\gamma_{rs}N_s]. \end{aligned} \tag{10.24}$$

The expansion of the exponential operator in Equation (10.24) will give a product of factors similar to the one which occurs in Equation (10.22). It follows from Equation (10.24) that within a manifold of states defined by a set of eigenvalues to the operators N_r, the compound operators of Equations (10.20) and (10.21) will remain basic for the treatment. They act as operators for elementary excitations for the case of separated atoms.

The algebraic steps in the former part of this section serve as a basis for the development in the following of a perturbation expansion of the evolution operator in a similar way as used by Bulaevskii. We will only consider terms of the second order in the perturbation operator,

$$H_{\text{pert}} = \sum \beta_{rs} a^{\dagger}_{r\nu} a_{s\nu}, \tag{10.25}$$

since there are no first order energy shifts, and degenerate second order perturbation theory is required to find the stable zero order states.

Let P be the projection operator unto the particular degenerate manifold of H_{atoms} in which we are interested and let it be specified by the eigenvalues of all N_r,

$$N_r P = q_r P. \tag{10.26}$$

The formal charge q_r of electrons on atom r also defines uniquely the unperturbed energy,

$$H_{\text{atoms}} P = E_0 P, \tag{10.27}$$

$$E_0 = \sum \left[\alpha_r q_r + \gamma_{rr} \binom{q_r}{2} \right] + \tfrac{1}{2} \sum{}' \gamma_{rs} q_r q_s. \tag{10.28}$$

We have already observed that there are no first order energy shifts and we can verify that

$$P H_{\text{pert}} P = 0. \tag{10.29}$$

Schrödinger perturbation theory leads us to study the second order reduced hamiltonian, which is given by

$$\begin{aligned} H_{\text{red}} &= P H_{\text{pert}} \left[\frac{1-P}{E_0 - H_{\text{atoms}}} \right] H_{\text{pert}} P \\ &= \lim_{\delta \to 0} \; -i \int_{-\infty}^{0} P H_{\text{pert}}(0) H_{\text{pert}}(t) P e^{\delta t} dt. \end{aligned} \tag{10.30}$$

The perturbation operator is obtained in the interaction representation by the insertion of the expression (10.24) in the form (10.25). We consider the integrand in Equation (10.30) to find that it has the form

$$P H_{\text{pert}}(0) H_{\text{pert}}(t) P = P \sum_{rs,\, \nu\nu'} \beta_{rs} a^{\dagger}_{r\nu} a_{s\nu} \beta_{sr} a^{\dagger}_{s\nu'}(t) a_{r\nu'}(t) P. \tag{10.31}$$

The time dependence is simplified if we take Equation (10.26) into account. Let us define

$$W_r = \alpha_r + \sum_{s \neq r} \gamma_{rs} q_s, \tag{10.32}$$

and observe that the commutation relation

$$[N_r, a_{r\nu}] = -a_{r\nu}$$

can be used to obtain

$$\exp[it N_r \gamma_{sr}]\, a_{r\nu} = a_{r\nu} \exp[it(N_r - 1)\, \gamma_{sr}].$$

After some further algebraic manipulations we find that

$$\begin{aligned}\Sigma_{\nu\nu'} P a^{\dagger}_{r\nu} a_{s\nu} a^{\dagger}_{s\nu'}(t) a_{r\nu'}(t) P= \\ = P\{\tfrac{1}{2}q_r(2-q_r)(2-q_s)(1-q_s) \exp [it(W_s-W_r-\gamma_{sr})] \\ +\tfrac{1}{2}q_r(q_r-1)(2-q_s)(1-q_s) \exp [it(W_s-W_r-\gamma_{sr}-\gamma_{rr})] \\ +\tfrac{1}{2}q_r(2-q_r)\, q_s(2-q_s) \exp [(W_s-W_r-\gamma_{ss}-\gamma_{rs})] \\ +\tfrac{1}{2}q_r(q_r-1)\, q_s(2-q_s) \exp [it(W_s-W_r+\gamma_{ss}-\gamma_{sr}-\gamma_{rr})] \\ -2\bar{S}_r.\bar{S}_s \exp [it(W_s-W_r+\gamma_{ss}-\gamma_{rs})]\}.\end{aligned} \tag{10.33}$$

The first four terms of the right hand side of Equation (10.33) will, when inserted into Equation (10.31) and (10.30) give a constant contribution to the reduced hamiltonian of Heisenberg's type. We have the relations

$$H_{\text{red}} = \text{constant} - \sum J_{rs}\, \bar{S}_r.\bar{S}_s, \tag{10.34}$$

$$J_{rs} = -|\beta_{rs}|^2 [(W_s-W_r+\gamma_{ss}-\gamma_{rs})^{-1} + (W_r-W_s+\gamma_{rr}-\gamma_{sr})^{-1}]. \tag{10.35}$$

A system with identical atoms and with one electron per atom is seen to behave in this approximation as a Heisenberg antiferromagnet with the exchange integrals

$$J_{rs} = -2\, |\beta_{rs}|^2/(\gamma_{rr}-\gamma_{rs}). \tag{10.36}$$

The constant is in this case such that we get

$$H_{\text{red}} = -\sum_{r,\,s} J_{rs}\,(-\tfrac{1}{4}+\bar{S}_r.\bar{S}_s), \tag{10.37}$$

a result which we will use later.

Our analysis above has demonstrated the nature of the spectrum of the hamiltonian in the Pariser–Parr–Pople model when the off-diagonal elements of the core-hamiltonian are small. The elementary excitation operators are essentially atomic and correspond to localized processes. There will be families of states that are almost degenerate and which are connected to one another through spin excitations.

Interaction of Atoms

This section deals with the calculation of double-time Green's functions of Chapter 5 with the view of studying the propagation of electrons through a system with small interactions between atoms.

The Green's functions are defined here with respect to a specific state, called the ground state, of the system rather than an ensemble as is used in statistical mechanics. Similarly it will follow that the decoupling procedure for the equations of motion will be referred to this state. The detailed nature of the state will ideally be the result of the calculation but it will become clear that its construction is a complicated matter. Ground state properties can be readily calculated without explicit knowledge of the state. There will, however, be important questions about the so-called N-representability problem that must be left unanswered in this presentation.

Two kinds of elementary excitation operators were introduced in Equations (10.20) and (10.21) in describing the limit of separated atoms. These operators will form the basis also for the discussion of interacting atoms.

Rather than working with them in the given form we will use linear combinations, which are given as

$$a_{r\nu} = (1 - n_{r-\nu})\, a_{r\nu} + n_{r-\nu} a_{r\nu}, \tag{10.38}$$

and

$$b_{r\nu} = z_r(n_{r-\nu} - \langle n_{r-\nu}\rangle)\, a_{r\nu}, \tag{10.39}$$

with

$$z_r = [\,\langle n_{r-\nu}\rangle\,(1 - \langle n_{r-\nu}\rangle)\,]^{-\frac{1}{2}}. \tag{10.40}$$

We assume here and in the following that all expectation values are calculated in a *singlet spin state*. When we think of the ground state as an expansion of states from the valence bond theory, we find that while $a_{r\nu}$ annihilates an electron on atom r with spin ν, $b_{r\nu}$ will have the same effect with a change of relative phase and weights of the structures for which atom r is "ionic" and "covalent".

The following relations are satisfied by these operators

$$\langle [b_{r\nu}, a^{\dagger}_{s\nu}]_{+}\rangle = 0, \tag{10.41}$$

$$\langle [b_{r\nu}, b^{\dagger}_{s\nu'}]_{+}\rangle = \delta_{rs}\delta_{\nu\nu'}. \tag{10.42}$$

There are two Green's functions in which we are interested, the single electron propagator $G_{rs}(t)$,

$$G_{rs}(t - t') = \langle\langle a_{r\nu}(t)\,;\, a^{\dagger}_{s\nu}(t')\rangle\rangle, \tag{10.43}$$

and the mixed $K_{rs}(t)$,

$$K_{rs}(t - t') = \langle\langle b_{r\nu}(t);\, a_{\dagger\nu}(t')\rangle\rangle. \tag{10.44}$$

They will be calculated in terms of their Fourier transforms,

$$G_{rs}(E) = \langle\langle a_{r\nu};\, a^{\dagger}_{s\nu}\rangle\rangle_E = \int G_{rs}(t)\, e^{itE} dt, \tag{10.45}$$

which satisfy the equations,

$$E\, G_{rs}(E) = \delta_{rs} + \langle\langle [a_{r\nu}, H];\, a^{\dagger}_{s\nu}\rangle\rangle_E, \tag{10.46}$$

and

$$E K_{rs}(E) = \langle\langle [b_{r\nu}, H];\, a^{\dagger}_{s\nu}\rangle\rangle_E, \tag{10.47}$$

as well as certain initial conditions. The hamiltonian H in these equations is the one given by Equation (10.12).

We wish to truncate the equation of motion of $G_{rs}(E)$ to a linear expression of G's and K's, and similarly for the equation of motion of $K_{sr}(E)$. Equivalently we formulate the problem as one of finding a linear expansion in terms of a's and b's of a general fermion operator expression X:

$$X=\sum (f_{rv}\, a_{rv}+I_{rv}\, b_{rv}). \tag{10.48}$$

The coefficients f_{rv} and I_{rv} are determined from the requirement that the effect of the operator X on the ground state should equal the effect of the expansion on the same. This gives the two equations

$$X\,|0\rangle=\sum (f_{rv}\, a_{rv}+I_{rv}\, b_{rv})|\,0\rangle \tag{10.49}$$

$$\langle 0\,|X=\sum \langle 0|(f_{rv}a_{rv}+I_{rv}b_{rv}), \tag{10.50}$$

which imply that

$$f_{rv}=\langle [X, a_{rv}^{\dagger}]_{+}\rangle, \tag{10.51}$$

and

$$I_{rv}=\langle [X, b_{rv}^{\dagger}]_{+}\rangle. \tag{10.52}$$

X will in our case be $[a_{rv}, H]$ and $[b_{rv}, H]$, which give us reason to define the following matrix elements:

$$f_{rs}=\langle [\,[a_{rv}, H], a_{sv}^{\dagger}]_{+}\rangle, \tag{10.53}$$

$$I_{rs}=\langle [\,[a_{rv}, H], b_{sv}^{\dagger}]_{+}\rangle, \tag{10.54}$$

$$\hat{f}_{rs}=\langle [\,[b_{rv}, H], b_{sv}^{\dagger}]_{+}\rangle. \tag{10.55}$$

Disregarding truncation errors from Equation (10.48) we obtain from Equations (10.46) and (10.47) the matrix equation

$$\begin{pmatrix} E\mathbf{1}-\mathbf{f} & -\mathbf{I} \\ -\mathbf{I}^{\dagger} & E\mathbf{1}-\hat{\mathbf{f}} \end{pmatrix} \begin{pmatrix} \mathbf{G}(E) \\ \mathbf{K}(E) \end{pmatrix} = \begin{pmatrix} \mathbf{1} \\ \mathbf{0} \end{pmatrix}. \tag{10.56}$$

The ordinary molecular orbital approximation according to the Hartree–Fock method is obtained from Equation (10.56) when the matrix $\mathbf{I}$ is neglected and there is no coupling between the Green's functions of G- and K-type.

The solution of Equation (10.56) is obtained by standard methods once the matrix elements are found. They involve the calculation of expectation values which can be found when the solution of Equation (10.56) is available and we are faced with a self-consistency requirement which is similar to but more complex than the one met with in the Hartree–Fock method. We leave the problem of calculating the matrix elements to the next section.

It is convenient to discuss the solution of Equation (10.56) in terms of the eigenvalues and eigenvectors obtained for the homogeneous equation system which is obtained when the right hand side is replaced by zero. We define these so that

$$\epsilon_k \begin{pmatrix} \mathbf{u}_k \\ \mathbf{v}_k \end{pmatrix} = \begin{pmatrix} \mathbf{f} & \mathbf{I} \\ \mathbf{I}^\dagger & \hat{\mathbf{f}} \end{pmatrix} \begin{pmatrix} \mathbf{u}_k \\ \mathbf{v}_k \end{pmatrix}, \tag{10.57}$$

and

$$\mathbf{u}_k^\dagger \mathbf{u}_{k'} + \mathbf{v}_k^\dagger \mathbf{v}_{k'} = \delta_{kk'}. \tag{10.58}$$

The eigenvectors serve to define elementary excitation operators $A_{k\nu}$ as

$$A_{k\nu} = \Sigma_r (u_{kr}^\dagger a_{r\nu} + v_{kr}^\dagger b_{r\nu}), \tag{10.59}$$

which can be used equivalently with the a's and b's.

$$a_{r\nu} = \Sigma\, u_{rk} A_{k\nu}, \tag{10.60}$$

$$b_{r\nu} = \Sigma\, v_{rk} A_{k\nu}. \tag{10.61}$$

Their importance depends upon the property that

$$\langle [\, [A_{k\nu}, H], A_{k'\nu'}^\dagger]_+ \rangle = \epsilon_k \delta_{kk'} \delta_{\nu\nu'}, \tag{10.62}$$

and that

$$(E - \epsilon_k) \langle\langle A_{k\nu}; A_{k'\nu}^\dagger \rangle\rangle_E = \delta_{kk'}. \tag{10.63}$$

It is necessary at this point to consider the previously mentioned boundary conditions. The behaviour of the Green's functions at the poles of their Fourier transforms determines uniquely the various properties of the solution and the only sensible specification with regard to the poles is to require that the inversion of the Fourier transform should involve an integration along the real axis in the complex E-plane such that all poles to the left of a point μ are circumvented below the axis and all poles to the right of μ are circumvented above the axis. Equivalently we can say that to each pole ϵ_k an infinitesimal part is added or subtracted when $\epsilon_k - \mu$ is negative or positive respectively:

$$\epsilon_k \to \epsilon_k + i\delta\, sgn\, (\mu - \epsilon_k), \qquad \delta > 0. \tag{10.64}$$

The choice of the point μ is critical and is given by the condition that the ground state should contain a certain number of electrons. It is not apparent that this number always will be an integer and we might have to accept states of the type used by Bardeen, Cooper, and Schrieffer in their theory of superconductivity, i.e. a superposition of states with varying number of particles. This difficulty has not occurred in the applications carried out so far.

The number of electrons in the ground state is given by the propagator as

$$N = \Sigma_{r\nu} \langle n_{r\nu} \rangle = 2\, \Sigma_r \lim_{t \to 0-} -iG_{rr}(t) = 2\, \Sigma_r (2\pi\, i)^{-1} \int_C dE\, G_{rr}(E). \tag{10.65}$$

The contour C in the complex E-plane is a closed contour which encircles the part of the real axis where $E < \mu$. Thus we get

$$N = 2 \sum_r \sum_{(k)} |u_{rk}|^2, \tag{10.66}$$

where the summation over k extends over the values for which $\epsilon_k < \mu$.

The number of electrons, N, is not necessarily given by an integer when Equation (10.66) is applied as noticed above but we can compute the difference from one as follows. We have the identity

$$n_{r\nu} = b^{\dagger}_{r\nu} b_{r\nu} + (2\langle n_{r-\nu}\rangle - 1)\, z_r a^{\dagger}_{r\nu} b_{r\nu}, \tag{10.67}$$

which leads us to the result that

$$N = 2 \sum_r \sum_{(k)} [\, |v_{rk}|^2 + (\langle N_r\rangle - 1) z_r u^{\dagger}_{kr} v_{rk}] \tag{10.68}$$

or by averaging

$$N = \sum_{(k)} \sum_r [\, |u_{rk}|^2 + |v_{rk}|^2 + (\langle N_r\rangle - 1)\, z_r u^{\dagger}_{kr} v_{rk}]. \tag{10.69}$$

The case with one electron per atom, $\langle N_r\rangle = 1$, gives then immediately an integer result from the normalization condition (10.58), but there might be a difference in other cases.

Alternative forms are also available for the calculation of the total energy in the ground state. The most straight-forward method is to compute the expectation value of the hamiltonian directly. Another common technique in the application of Green's functions makes use of the identity

$$\sum_{r\nu} a^{\dagger}_{r\nu} [a_{r\nu}, H] = \sum (\alpha_r \delta_{rs} + \beta_{rs})\, a^{\dagger}_{r\nu} a_{s\nu} + \sum{}' \gamma_{rs} n_{r\nu} n_{s\nu'}, \tag{10.70}$$

so that we find

$$\begin{aligned} E_0 = \langle H\rangle &= (2\pi i)^{-1} \int_C dE |\sum_{rs} [(\alpha_r + E)\, \delta_{rs} + \beta_{rs}]\, G_{sr}(E) \\ &= \sum_{rs} \sum_{(k)} [(\alpha_r + \epsilon_k)\, \delta_{rs} + \beta_{rs}]\, u_{sk} u^{\dagger}_{kr}. \end{aligned} \tag{10.71}$$

The form (10.71) is of course also valid in the Hartree–Fock approximation where all v_{kr}'s vanish, but in the more general case it might not agree with the value obtained from the direct calculation of the expectation value of the hamiltonian. This is another difficulty occurring in the Green's function method which ideally should be taken into account by some auxiliary condition on the decoupling procedure.

We might also look upon the consistency difficulties of the previous paragraphs as connected with the preservation of certain formal equivalencies in an approximate theory. An obvious relation following from the Hellmann–Feynman theorem is the expression for the bond order as a derivative of the total energy,

$$p_{sr}=\tfrac{1}{2}\,\partial E_0/\partial\beta_{rs}=\tfrac{1}{2}\sum_\nu \langle a^\dagger_{r\nu}a_{s\nu}+a^\dagger_{s\nu}a_{r\nu}\rangle$$
$$=(\pi i)^{-1}\int_C dE\; G_{sr}(E). \tag{10.72}$$

The simultaneous validity of Equations (10.71) and (10.72) puts restrictions on the relations of different expectation values and thereby on the decoupling procedure. There is no simple way of incorporating these conditions and it is not done in the present development.

Our final consideration in this section will be the comparison of Equation (10.72) and the energy computed from Equation (10.37). The bond order should be computed to order β_{rs} and we can classify the matrix elements of Equation (10.56) as follows when we assume that $\langle N_r\rangle=1$ and all atoms are equivalent

$$f_{rr}=W+\mathcal{O}(\beta^2), \tag{10.73}$$
$$f_{rs}=\beta_{rs}-\tfrac{1}{2}\gamma_{rs}\,p_{rs}, \tag{10.74}$$
$$I_{rr}=\tfrac{1}{2}\gamma_{rr}+\mathcal{O}\,(\beta^2), \tag{10.75}$$
$$I_{rs}=\mathcal{O}\,(\beta^2), \tag{10.76}$$
$$\hat{f}_{rr}=W+\mathcal{O}\,(\beta^2), \tag{10.77}$$
$$\hat{f}_{rs}=\beta_{rs}\,\langle[\,[b_{r\nu},\,a^\dagger_{r\nu}\,a_{s\nu}+a^\dagger_{r-\nu}\,a_{s-\nu}+a^\dagger_{s-\nu}a_{r-\nu}],\,b^\dagger_{s\nu}]_+\rangle+\tfrac{1}{2}\gamma_{rs}\,p_{rs}+\mathcal{O}\,(\beta^2). \tag{10.78}$$

It is a straight-forward perturbation calculation on Equation (10.56) to obtain the Green's function to order β and to evaluate the contour integral, μ can be chosen equal to W, which gives

$$p_{rs}=[\hat{f}_{rs}-f_{rs}]/\gamma_{rr}. \tag{10.79}$$

This is an implicit equation for p_{rs} according to Equations (10.74) and (10.78) which can be solved to give

$$p_{rs}=-[\,\beta_{rs}/(\gamma_{rr}-\gamma_{rs})]\,[1-4\,\langle\bar{S}_r.\bar{S}_s\rangle], \tag{10.80}$$

when terms of higher order in β are neglected in the evaluation of the expectation value in Equation (10.78) and spin symmetry is used.

Thus we obtain here the correct result provided we can calculate the expectation value of the product of spin operators. The analysis of this problem is given in the following section.

Calculation of Expectation Values

The matrix elements defined by Equation (10.53)–(10.55) will be calculated in detail in this section under the assumption that a ground state exists such that we can determine the Green's functions in the manner described in the previous part of the chapter. This assumption takes the form that

$$\langle A^{\dagger}_{k\nu} A_{k'\nu'} \rangle = \nu_k \delta_{kk'} \delta_{\nu\nu'}, \tag{10.81}$$

where the "occupation number" ν_k is determined such that

$$\nu_k = \begin{cases} 1 & \text{if } \epsilon_k < \epsilon_f \\ 0 & \text{if } \epsilon_k > \epsilon_f. \end{cases} \tag{10.82}$$

The implication of this is the property

$$[(1-\nu_k) A_{k\nu} + \nu_k A^{\dagger}_{k\nu}] \, |E_0\rangle = 0. \tag{10.83}$$

It has not been proven that the conditions imposed on the ground state by Equation (10.83) are incompatible and the validity of this property is basic for the further calculations.

All expectation values with which we are concerned can be written in the form $\langle X^{\dagger} a_{r\nu} \rangle$ in which X is a fermion-like annihilation operator. We employ the expansion (10.60) and the property (10.83) in order to obtain

$$\begin{aligned} \langle X^{\dagger} a_{r\nu} \rangle &= \Sigma_k u_{rk} \nu_k \langle X^{\dagger} A_{k\nu} \rangle = \Sigma_k u_{rk} \nu_k \langle [X^{\dagger}, A_{k\nu}]_+ \rangle \\ &= \Sigma_{ks} u_{rk} \nu_k u^{\dagger}_{ks} \langle [X^{\dagger}, a_{s\nu}]_+ \rangle + \Sigma_{ks} u_{rk} \nu_k v^{\dagger}_{ks} \langle [X^{\dagger}, b_{s\nu}]_+ \rangle. \end{aligned} \tag{10.84}$$

It will be expedient to introduce the notations

$$g_{rs} = (2\pi i)^{-1} \int_C dE \, G_{rs}(E) = \Sigma_k u_{rk} \nu_k u^{\dagger}_{ks} = g_{sr}, \tag{10.85}$$

$$k_{rs} = (2\pi i)^{-1} \int_C dE \, K_{rs}(E) = \Sigma_k v_{rk} \nu_k u^{\dagger}_{ks} \tag{10.86}$$

where it is assumed that the ground state is non-degenerate and all g's and k's are real. Thus we get from Equation (10.78) that

$$\langle X^{\dagger} a_{r\nu} \rangle = \Sigma_s g_{rs} \langle [X^{\dagger}, a_{s\nu}]_+ \rangle + \Sigma_s \langle [X^{\dagger}, b_{s\nu}]_+ \rangle \, k^{*}_{sr}. \tag{10.87}$$

Equation (10.81) is generally an implicit relation which leads to an equation system for several expectation values. Only when X is a simple fermion or quasi-fermion operator do we get the immediate result:

$$\langle a^{\dagger}_{s\nu} a_{r\nu} \rangle = g_{rs}, \tag{10.88}$$

and

$$\langle b^{\dagger}_{s\nu} a_{r\nu} \rangle = k^{*}_{sr} \tag{10.89}$$

in accordance with the initial value conditions on the Green's functions.

There is formally exact agreement between the matrix elements f_{rs} and the Fock operator elements in Hartree–Fock theory. It holds that

$$f_{rs} = \delta_{rs} [\alpha_r + 2 \Sigma_{r'} \gamma_{rr'} g_{r'r'}] + \beta_{rs} - \gamma_{rs} g_{rs}. \tag{10.90}$$

The difference lies in the calculation of the g_{rs}'s.

We consider the elements of the I-matrix next. They are evaluated from the form

$$I_{rs}=\delta_{rs}\sum_{r'\nu'}\gamma_{rr'}z_r\langle(n_{r-\nu}-g_{rr})\,n_{r'\nu'}\rangle-\gamma_{rs}k^*_{sr}. \tag{10.91}$$

The application of Equation (10.87) to $\langle(n_{r\nu}-g_{rr})\,n_{r'\nu'}\rangle$ leads to a system of equations,

$$\langle(n_{r\nu}-g_{rr})\,n_{r'\nu'}\rangle=\delta_{\nu\nu'}\,[g_{rr}\delta_{rr'}-g^2_{rr'}-|k_{rr'}|^2]+k_{r'r'}z_{r'}\langle(n_{r\nu}-g_{rr})\,n_{r'-\nu'}\rangle \tag{10.92}$$

with the solution for the interesting quantity,

$$\langle(n_{r\nu}-g_{rr})\,(n_{r'\nu'}+n_{r'-\nu'})\rangle=[g_{rr}\delta_{rr'}-g^2_{rr'}-|k_{rr'}|^2]/[1-k_{r'r'}z_{r'}]. \tag{10.93}$$

The explicit expression for the matrix element I_{rs} arises from insertion of the expression (10.93) in Equation (10.91) and will be omitted here. Matrix elements of the type $\hat{f}_{rs}$ are more complicated and we will first rewrite them in a less compact form. It follows from Equation (10.55) and the definition (10.39) that

$$\begin{aligned}\hat{f}_{rs}&=z_r\langle(n_{r-\nu}-g_{rr})\,[\,[a_{r\nu},H],b^\dagger_{s\nu}]_++[n_{r-\nu},H]\,[a_{r\nu},b^\dagger{}_\nu]_++[b^\dagger_{s\nu},[n_{r-\nu},H]\,]\,a_{r\nu}\rangle\\&=z_rz_s\langle(n_{r-\nu}-g_{rr})\,(n_{s-\nu}-g_{ss})\,[\,[a_{r\nu},H],a^\dagger_{s\nu}]_++\delta_{rs}[n_{r-\nu},H]\,(n_{r-\nu}-g_{rr})\\&\qquad+a^\dagger_{s\nu}[n_{s-\nu},[n_{r-\nu},H]\,]\,a_{r\nu}\rangle.\end{aligned} \tag{10.94}$$

The diagonal elements are

$$\hat{f}_{rr}=f_{rr}+z_r\,(1-2g_{rr})\,I_{rr}+z_r\sum_{r'}(\beta_{rr'}k^*_{r'r}+k_{rr'}\beta_{r'r})+z^2_r\,(2g_{rr}-1)\sum_{r'}g_{rr'}\beta_{r'r}. \tag{10.95}$$

We will show that for a cyclic even alternant hydrocarbon it is a self-consistent assumption to let the last three terms vanish.

Off-diagonal elements are given by

$$\hat{f}_{rs}=\beta_{rs}z_rz_s\langle(n_{r-\nu}-g_{rr})\,(n_{s-\nu}-g_{ss})+a^\dagger_{s\nu}a_{s-\nu}a^\dagger_{r-\nu}a_{r\nu}-a^\dagger_{s\nu}a^\dagger_{s-\nu}a_{r-\nu}a_{r\nu}\rangle-\gamma_{rs}\langle b^\dagger_{s\nu}b_{r\nu}\rangle. \tag{10.96}$$

The last term is evaluated thus:

$$\langle b^\dagger_{s\nu}b_{r\nu}\rangle=\sum v_{rk}\nu_k\langle[b^\dagger_{s\nu},A_{k\nu}]_+\rangle=\sum v_{rk}\nu_kv^\dagger_{ks}=\hat{g}_{rs}. \tag{10.97}$$

We find the first part of the first term from Equation (10.92) as

$$\langle(n_{r-\nu}-g_{rr})\,(n_{s-\nu}-g_{ss})\rangle=[\delta_{sr}g_{rr}-g^2_{rs}-|k_{rs}|^2]/[1-k^2_{ss}z^2_s], \tag{10.98}$$

a form which is not symmetrically written in terms of r and s.

Similarly we derive

$$\langle a^\dagger_{s\nu} a_{s-\nu} a^\dagger_{r-\nu} a_{r\nu} \rangle = [\delta_{sr} g_{rr} - g^2_{rs} - |k_{rs}|^2]/[1 + k_{ss} z_s], \qquad (10.99)$$

and

$$\langle a^\dagger_{s\nu} a^\dagger_{s-\nu} a_{r-\nu} a_{r\nu} \rangle = [g^2_{rs} - k^{*2}_{r\,s} + k^*_{rs} g_{sr} z_r (1 - 2g_{rr})]/[1 - k_{ss} z_s]. \qquad (10.100)$$

It should be noticed that the quantity $k_{ss}z_s$ which appears frequently is a measure of the dispersion of the operator N_s:

$$k_{ss} z_s = z^2_s \langle (n_{s-\nu} - g_{ss})\, n_{s\nu} \rangle = \tfrac{1}{2} z^2_s \langle (N_s - \langle N_s \rangle)^2 \rangle - 1. \qquad (10.101)$$

When the coupling between atoms is small there will be almost vanishing denominators in the expressions (10.98) and (10.99) and we need to consider the limit carefully. A study leads to the result that the limit can be given in terms of the spin operators of atoms r and s as was indicated above.

Application to Linear Chains

A linear chain is here taken to mean a system to which the hamiltonian (10.12) applies and where it is appropriate to assume periodicity in the arrangement of the atoms. In particular we can think of the annulenes.

Hückel theory for the even alternant hydrocarbons leads to the Coulson–Rushbrooke theorem and some other characteristic results which McLachlan has shown are valid also in the Pariser–Parr–Pople model. These are the well-known pairing relations between electronic states of alternant hydrocarbon cat- and anions. This particle hole symmetry is analogous to the situation discussed in Chapter 8 for electrons and holes in atomic subshells.

For even alternant hydrocarbons we can divide the atomic sites into two classes (even and odd) such that atoms belonging to the even class have nearest neighbors belonging to the odd class only, and vice versa. This symmetry property of the Pariser–Parr–Pople hamiltonian can be readily explored if we write

$$H(\mathrm{PPP}) = K\, N_{\mathrm{op}} + \Sigma'\, \beta_{rs} a^\dagger_{r\nu} a_{s\nu} + \Sigma_{r<s}\, \gamma_{rs}\, [(N_r - 1)(N_s - 1) - 1] + \Sigma\, \gamma_{rr} [(n_{r+} - \tfrac{1}{2})(n_{r-} - \tfrac{1}{2}) - \tfrac{1}{4}] \qquad (10.102)$$

where $K = \alpha_r + \frac{1}{2}\gamma_{rr} + \Sigma_{s \neq r} \gamma_{rs}$ is a constant for all atoms r.

Consider the complete particle-hole transformation by the unitary operator $U = \exp(iS)$, where

$$S = S^\dagger = (\pi/2) \left[\sum^{\mathrm{even}} (a^\dagger_{r+} a^\dagger_{r-} + a_{r-} a_{r+}) - \sum^{\mathrm{odd}} (a^\dagger_{r+} a^\dagger_{r-} + a_{r-} a_{r+}) \right].$$

From the results in Chapter 8 we obtain the following relations for the transformed field operators $\hat{a} = U a U^\dagger$:

$$\begin{aligned} \hat{a}_{r+} &= -i a^\dagger_{r-} \\ \hat{a}_{r-} &= i a^\dagger_{r+} \end{aligned} \qquad \text{for } r \text{ even} \qquad (10.103)$$

and

$$\begin{aligned}\hat{a}_{r+} &= i a_{r-}^{\dagger} \\ \hat{a}_{r-} &= -i a_{r+}^{\dagger}\end{aligned} \quad \text{for } r \text{ odd.} \tag{10.104}$$

It should be observed that when only nearest neighbor interactions are considered for β_{rs}, the atomic sites r and s belong to different classes and the associated "bond order" operator product satisfies the relation

$$\hat{a}_{r\nu}^{\dagger}\hat{a}_{s\nu} = a_{s\nu}^{\dagger}a_{r\nu}. \tag{10.105}$$

The transformed hamiltonian then becomes

$$\hat{H}(\text{PPP}) = H(\text{PPP}) + 2K(n - N_{\text{op}}), \tag{10.106}$$

where n is the number of carbon atoms and we have considered β_{rs} to be real, since no magnetic fields are included in this treatment.

The implications of these results for the present study is a number of conditions on the matrix elements g_{rs}, k_{rs} and $\hat{g}_{rs}$ which we will show are self-consistent, when the assumptions for an alternant system are invoked. Table 9 lists the important consequences of the alternant character of the model.

TABLE 9. Matrix Elements for a Cyclic Alternant Hydrocarbon in the Pariser–Parr–Pople Model

Matrix element	r, s both odd or even	r, s not both odd or even
β_{rs}	0	$\neq 0$
g_{rs}, $\hat{g}_{rs}$	$\frac{1}{2}\delta_{rs}$	$\neq 0$
k_{rs}	x	0
f_{rs}	$\mu\delta_{rs}$	$\neq 0$
I_{rs}	x	0
$\hat{f}_{rs}$	$\mu\delta_{rs}$	$\neq 0$

The parameter μ can be chosen equal to the common diagonal element of matrices $\mathbf{f}$ and $\hat{\mathbf{f}}$ since the resulting secular determinant of Equation (10.56) will show pairing of roots around μ and the number of roots equals twice the number of electrons (cf. Equation (10.69)). The pairing property is shown from Equation (10.57) by use of the transformations

$$\begin{aligned}\epsilon_k - \mu &\rightarrow \mu - \bar{\epsilon}_{-k}, \\ u_{rk} &\rightarrow (-)^r u_{r-k}, \\ v_{rk} &\rightarrow -(-)^r v_{r-k},\end{aligned} \tag{10.107}$$

which also can be used to demonstrate the self-consistency of the solution. An obvious feature of the pairing will be the existence of a "gap" in the energy spectrum around μ. This does not appear to be an unreasonable situation, since even Hartree–Fock theory gives a very small number of states in the vicinity of the middle of the spectrum and all unrestricted Hartree–Fock procedures also give a finite "gap" for nonvanishing interactions in a linear system.

Electronic interaction integrals enter the various matrix elements but the most significant contribution from them comes into the matrix **I**, particularly its diagonal elements. We preserve most of the characteristics of the problem by omitting off-diagonal parts of **I** or equivalently to put all two-center integrals γ_{rs} to zero. This is not such a crude approximation as it may seem since those integrals occur only in conjunction with small factors everywhere except in diagonal elements f_{rr}, which anyway are put equal to a reference point on the energy scale, and the nearest neighbor elements f_{rs} and $\hat{f}_{rs}$ where we might "renormalize" the β-value to account for the last term in Equation (10.90). These approximations reduce the problem considerably and there remains the following matrix elements when $\mu=0$,

$$
\begin{aligned}
f_{rs} &= \begin{cases} \beta & r, s \text{ neighbors,} \\ 0 & \text{otherwise,} \end{cases} \\
I_{rs} &= I\delta_{rs} = \tfrac{1}{2}\gamma_{rr}\delta_{rs}, \\
\hat{f}_{rs} &= \begin{cases} \lambda\beta & r, s \text{ neighbors,} \\ 0 & \text{otherwise,} \end{cases} \\
\lambda &= -12\, g_{rs}^{2}/(1-4k_{ss}^{2}), \qquad r, s \text{ neighbors.}
\end{aligned}
\tag{10.108}
$$

The solution of the eigenvalue problem is very simply obtained from the *ansatz* for the eigenvectors,

$$
\begin{aligned}
\sqrt{N}u_{rk} &= \cos\theta \exp(i\phi r), \\
\sqrt{N}v_{rk} &= \sin\theta \exp(i\phi r),
\end{aligned}
\tag{10.109}
$$

with the boundary conditions for a periodic solution:

$$
\begin{aligned}
u_{rk} &= u_{r+Nk}, \\
v_{rk} &= v_{r+Nk}.
\end{aligned}
\tag{10.110}
$$

Thus we get the equation system

$$
\begin{aligned}
(\epsilon_k - 2\beta\cos\phi)\cos\theta &= I\sin\theta, \\
(\epsilon_k - 2\lambda\beta\cos\phi)\sin\theta &= I\cos\theta.
\end{aligned}
\tag{10.111}
$$

Two branches of eigenvalues arise from Equation (10.111) and we will use the boundary conditions (10.110), in accordance with the pairing property

(10.107) to define

$$\begin{aligned}\epsilon_k &= (1+\lambda)\,\beta\cos(2\pi k/N)+[I^2+(1-\lambda)^2\beta^2\cos^2(2\pi k/N)]^{\frac{1}{2}},\\ k &= 0, 1, 2, \dots N-1,\\ \epsilon_k &= -\epsilon_{|k|},\ k=-N,\ -N+1,\dots -2,\ -1.\end{aligned} \tag{10.112}$$

The eigenvectors can now be determined in terms of parameters θ_k from Equation (10.100) and we are then ready to calculate the parameter λ self-consistently.

We derive first that

$$\begin{aligned}Ng_{01} &= -\sum_{k=-N}^{-1}\cos^2\theta_k\cos(2\pi k/N)\\ &= -\tfrac{1}{2}(1-\lambda)(\beta/I)\sum_{k=0}^{N-1}\cos^2(2\pi k/N)\,[1+(1-\lambda)^2(\beta/I)^2\cos^2(2\pi k/N)]\end{aligned} \tag{10.113}$$

which gives us the two limiting results that

$$g_{01} \underset{\beta/I\to 0}{\to} -(1-\lambda)(\beta/4I) \tag{10.114}$$

and

$$g_{01} \underset{I/\beta\to 0}{\to} \sum_{k=0}^{N-1}|\cos(2\pi k/N)|/2N \underset{N\to\infty}{\to} 1/\pi. \tag{10.115}$$

Secondly we find that

$$\begin{aligned}N\,k_{00} &= \sum_{k=-N}^{-1}\cos\theta_k\sin\theta_k\\ &= -\tfrac{1}{2}\sum_{k=0}^{N-1}[1+(1-\lambda)^2(\beta/I)^2\cos^2(2\pi k/N)]^{-\frac{1}{2}},\end{aligned} \tag{10.116}$$

with limiting values

$$k_{00} \underset{\beta/I\to 0}{\to} -\tfrac{1}{2}+\tfrac{1}{2}(1-\lambda)^2(\beta/2I)^2, \tag{10.117}$$

and

$$k_{00} \underset{I/\beta\to 0}{\to} 0. \tag{10.118}$$

We conclude that when $N\geq 6$ it holds that

$$-\tfrac{3}{2}<\lambda<-12/\pi^2=-1{\cdot}216 \tag{10.119}$$

and that it is rather insensitive to the ratio β/I. The two branches of the eigenvalue spectrum defined by Equation (10.111) form a hyperbola when the eigenvalue (10.112) are plotted against the Hückel energy $2\beta\cos(2\pi k/N)$ with asymptotes, the slope of which are unity and λ.

In order to compare the present theory with known results we have calculated the total energy of the linear chain in the limit of small β. The total energy of the Green's function method is obtained in two ways, from Equation (10.71) and also from the integral of Equation (10.72) as before. These results are compared with the one- and many-parameter alternant molecular orbital energies of Pauncz, de Heer and Löwdin, and with Hulthén's exact calculation for the antiferromagnetic linear chain in Table 10. It is noticed that Equation (10.71) gives an energy which is too low, a result which we interpret as a failure for the identity (10.70) within the present approximate calculation.

TABLE 10. Comparison of Calculations of the Energy of an Infinite Linear Chain in the Limit $|\beta/I| \ll 1$

Method	$E_0 I/2\beta^2 N$	% of exact value
One-parameter AMO	−0·4053	58·5
Many-parameter AMO	−0·5000	72·1
Equation (10.72)	−0·6250	90·2
Equation (10.71)	−0·8594	124·0
Hulthén	−0·6931	100·0

Notes and Bibliography

We have indicated one possible method to extend the study of Green's functions beyond the Hartree–Fock approximation and succeeded following the work of J. Hubbard in *Proc. Roy. Soc.* (1963), **A276**, 238; (1964), **A277**, 237, and **A281**, 401; (1965), **A285**, 542; (1967), **A296**, 82, and **A296**, 100 to obtain a solution with asymptotically correct properties for vanishing interactions between atoms. This chapter follows closely earlier work by the authors from 1967 in *Chem. Phys. Letters* **1**, 295, and from 1968 in *J. Chem. Phys.* **49**, 716.

Useful relations between matrix elements in the Pariser–Parr–Pople model have been obtained by Linderberg (1967) in *Chem. Phys. Letters*, **1**, 39, using field operator formalisms. The original works from which the name of the model stems are from 1953 by R. Pariser and R. G. Parr in *J. Chem. Phys.* **21**, 446 and 767, and by J. A. Pople in *Trans. Faraday Soc.* **42**, 1375. Extensive work on this model has been carried out by many, e.g. J. Koutecky (1967) in *J. Chem. Phys.* **47**, 1501. The superposition of configuration method by Cizek referred to in the text is from (1966) and published in *J. Chem. Phys.* **45**, 4256. P. G. Lykos and R. G. Parr (1956) discussed the so-called sigma-pi separation in *J. Chem. Phys.* **24**, 1166. E. A. Taft and H. R. Philipp (1965) published in *Phys. Rev.* **138**, A197, measurements and theoretical considerations on the optical reflectivity of graphite which give support to the ideas of sigma-pi separation and the simple ideas of screening discussed in this chapter.

The Hellmann–Feynman theorem used in developing the formula for the bond order as a derivative of the total energy has its origin in work by G. Hellmann (1937), "Einführung in die Quantenchemie", Leipzig, page 285, and in work by R. P. Feynman from 1939 published in *Phys. Rev.* **56**, 340.

The theorem by C. A. Coulson and G. S. Rushbrooke regarding even alternant hydrocarbons within the Hückel model was published in 1940 in *Proc. Camb. Phil.*

Soc. **40**, 193, and the pairing relations for the Pariser–Parr–Pople model were obtained by A. D. McLachlan (1959) in *Mol. Phys.* **2**, 271.

In the application to linear chains we have compared results with those obtained by R. Pauncz, J. de Heer and P. O. Löwdin (1962) using the alternant molecular orbital model as presented in *J. Chem. Phys.* **36**, 2247 and 2257.

Already in 1938 L. Hulthén did the exact calculation for the antiferromagnetic linear chain published in *Arkiv. Mat. Fys. Astr.* **26A**, no. 11.

Detailed calculations for the six-membered ring, using the approximation presented in this chapter, have been presented by J. Linderberg and E. W. Thulstrup (1968) in *J. Chem. Phys.* **49**, 710.

The linear chain problem has been solved by E. H. Lieb and F. Y. Wu in 1968, *Phys. Rev. Lett.* **20**, 1445, for the simplified hamiltonian with $\gamma_{rs}=\gamma\delta_{rs}$.

CHAPTER 11

Diagrammatic Expansions. Temperature Dependent Perturbation Theory

Partition Function

This chapter deals with the development of perturbation theory expansions of the partition function, the free energy, the electron propagator, and the polarization propagator. We find it natural to begin with the partition function,

$$Z(\beta)=Tr\exp[-\beta(H-\mu N_{op})], \tag{11.1}$$

and we consider the hamiltonian H as divided into an unperturbed part H_0 and a perturbation V. It is convenient at instances to assume that H_0 is a one-particle operator and to employ the basis set in which it has a diagonal form,

$$H_0=\sum_k \epsilon_k a_k^\dagger a_k. \tag{11.2}$$

We factor out, in the density operator, an unperturbed part,

$$\exp[-\beta(H_0-\mu N_{op})],$$

and a part which describes the effect of the perturbation,

$$S(\beta)=\exp[\beta(H_0-\mu N_{op})]\exp[-\beta(H-\mu N_{op})]. \tag{11.3}$$

The partition function for the unperturbed part is

$$Z_0(\beta)=Tr\exp[-\beta(H_0-\mu N_{op})]=\prod_k[1+\exp[-\beta(\epsilon_k-\mu)]], \tag{11.4}$$

and the total partition function is then

$$Z(\beta)=Z_0(\beta)\langle S(\beta)\rangle_0, \tag{11.5}$$

where $\langle\rangle_0$ indicates the average value formation over the unperturbed ensemble.

The operator $S(\beta)$ equals the unit operator for β equal to zero and we can then derive from Equation (11.3) that

$$S(\beta)=1+\int_0^\beta \frac{\partial S}{\partial\tau}d\tau=1-\int_0^\beta V(\tau)S(\tau)d\tau. \tag{11.6}$$

Here we have expressed the perturbation operator in the so-called interaction representation,

$$V(\tau)=\exp\,[\tau(H_0-\mu N_{\mathrm{op}})]\;V\exp\,[-\tau(H_0-\mu N_{\mathrm{op}})]. \tag{11.7}$$

We obtain by iteration of Equation (11.6) that

$$\begin{aligned}\langle S(\beta)\rangle_0 &= 1+\sum_{n=1}^{\infty}(-)^n\int_0^{\beta}d\tau_1\int_0^{\tau_1}d\tau_2\ldots\int_0^{\tau_{n-1}}d\tau_n\,\langle V(\tau_1)\,V(\tau_2)\ldots V(\tau_n)\rangle_0\\ &= 1+\sum_{n=1}^{\infty}(-)^n\,(n!)^{-1}\int_0^{\beta}\ldots\int_0^{\beta}d\tau_1\ldots d\tau_n\,\langle T[V(\tau_1)\ldots V(\tau_n)]\rangle_0.\end{aligned} \tag{11.8}$$

The "time-ordering" operator T arranges all operators so that the argument τ of the various operators in a product will be increasing when proceeding from right to left. It will also account for the necessary sign changes due to the anticommutation rules of fermion operators.

A rational approach to the evaluation of the terms in the series (11.8) is offered by the diagrammatic analysis introduced by Feynman for time-dependent problems and by Matsubara for the present case. Such a development is derived with the use of Wick's theorem.

We specify the perturbation operator V as having a one-particle and a two-particle contribution,

$$V=V_1+V_2, \tag{11.9}$$

as would be the case when the unperturbed hamiltonian is the Hartree–Fock effective operator. Our study will be concerned with the cases where we can employ the following representations of the operators:

$$V_1=\int dx\,dx'\,\psi^{\dagger}(x')\,v(x',x)\psi(x), \tag{11.10}$$

$$V_2=\tfrac{1}{2}\int dx\,dy\,\psi^{\dagger}(x)\psi^{\dagger}(y)w(x,y)\psi(y)\psi(x). \tag{11.11}$$

The similarity transformation (11.7) will affect the field operators so that we get

$$\begin{aligned}\psi(x\tau)&=\textstyle\sum_k u_k(x)\,a_k(\tau)\\ &=\textstyle\sum_k u_k(x)\,a_k\exp\,[-\tau\,(\epsilon_k-\mu)],\end{aligned} \tag{11.12}$$

and

$$\begin{aligned}\psi^{\dagger}(x\tau)&=\textstyle\sum_k u_k^{*}(x)a_k^{\dagger}(\tau)\\ &=\textstyle\sum_k u_k^{*}(x)a_k^{\dagger}\exp\,[\tau(\epsilon_k-\mu)].\end{aligned} \tag{11.13}$$

We prefer to use the notation $\psi^{\dagger}(x\tau)$ in Equation (11.13) although it is not the adjoint of $\psi(x\tau)$. It is convenient to define the "time-ordering" operation for the field operators so that for equal "times" the operators occur in the normal order where annihilation operators appear to the right of creation operators. Thus we can write

$$V_1(\tau) = -\int dx\, dx' v(x', x)\, T[\psi(x\tau)\psi^\dagger(x'\tau)] \tag{11.14}$$

$$V_2(\tau) = \tfrac{1}{2}\int dx\, dy\, w(x, y)\, T[\psi(x\tau)\psi(y\tau)\psi^\dagger(y\tau)\psi^\dagger(x\tau)]. \tag{11.15}$$

The single perturbation series (11.8) can now be expressed as a double sum when each factor $V(\tau)$ is a binomial $V_1(\tau) + V_2(\tau)$,

$$\begin{aligned}\langle S(\beta)\rangle_0 = \sum_{n=0}^{\infty}\sum_{m=0}^{\infty} (-)^{n+m}\, [n!m!]^{-1} \\ \times \int_0^\beta \cdots \int_0^\beta d\tau_1 \ldots d\tau_n d\tau_{n+1} \ldots d\tau_{n+m} \\ \times \langle T[V_1(\tau_1) \ldots V_1(\tau_n) V_2(\tau_{n+1}) \ldots V_2(\tau_{n+m})]\rangle_0,\end{aligned} \tag{11.16}$$

where the term for $n = m = 0$ is unity. A general term in this series is more explicitly written as

$$\begin{aligned}&\langle T[V_1(\tau_1) \ldots V_1(\tau_n) V_2(\tau_{n+1}) \ldots V_2(\tau_{n+m})]\rangle_0 \\ &= (-)^n \int dx_1 dx_1' v(x_1', x_1) \int dx_2 dx_2' v(x_2', x_2) \ldots \int dx_n dx_n' v(x_n', x_n) \\ &\times (\tfrac{1}{2})^m \int dx_{n+1} dy_{n+1} w(x_{n+1}, y_{n+1}) \ldots \int dx_{n+m} dy_{n+m} w(x_{n+m}, y_{n+m}) \\ &\times \langle T[\psi(x_1\tau_1) \ldots \psi(x_n\tau_n)\psi(x_{n+1}\tau_{n+1})\psi(y_{n+1}\tau_{n+1}) \ldots \psi(x_{n+m}\tau_{n+m}) \\ &\times \psi(y_{n+m}\tau_{n+m})\psi^\dagger(y_{n+m}\tau_{n+m})\psi^\dagger(x_{n+m}\tau_{n+m}) \ldots \psi^\dagger(y_{n+1}\tau_{n+1}) \\ &\times \psi^\dagger(x_{n+1}\tau_{n+1})\psi^\dagger(x_n'\tau_n) \ldots \psi^\dagger(x_1'\tau_1)]\rangle_0.\end{aligned} \tag{11.17}$$

Wick's theorem, which is proven in Appendix II p. 127, permits the evaluation of the expectation value above in the form of a determinant of contractions,

$$g(x\tau, x'\tau') = -\langle T[\psi(x\tau)\psi^\dagger(x'\tau')]\rangle_0, \tag{11.18}$$

and to analyse the various contributions we will now make use of diagrams.

Integrals like those in Equations (11.16) and (11.17) are performed in space and "time" and a graphical representation of the integrands is obtained by picturing the integration variables as points $(x\tau)$ in a plane. These points are joined by various elements representing the potentials $v(x', x)$ and $w(x, y)$ and contractions $g(x\tau, x'\tau')$. The latter have a directional property in that the point $(x\tau)$ is associated with an annihilation operator while the other argument occurs in a creation operator. Hence it will be drawn as an

TABLE 11. Diagrams for Potentials and Contractions

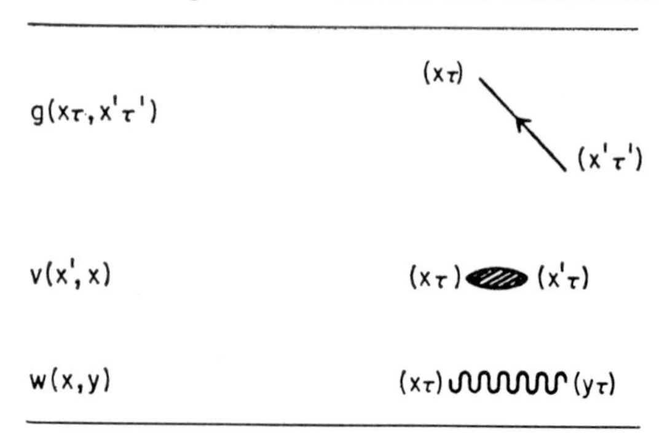

arrowed line from $(x'\tau')$ to $(x\tau)$. This and other constituents of the diagrams are displayed in Table 11.

Rules for the construction of diagrams are established from an inspection of the formula (11.17). Each factor $v(x_j', x_j)$ is put in through its symbol at its "time" τ_j and similarly each factor $w(x_k, y_k)$ appears at its "time" τ_k. Contraction symbols are then drawn from the corresponding points or vertices such that whenever an argument is associated with a creation operator in the perturbation term there will be a line leaving and whenever an argument occurs in conjunction with an annihilation operator a line will terminate at this vertex. A symbol for $v(x', x)$ will then occur with a line ending at x and one beginning at x'. The symbol for $w(x, y)$ will have lines entering and leaving at both x and y.

A general term of order n in V_1 and m in V_2 of Equation (11.16) will contain $(n+2m)!$ parts as is apparent from the determinantal form of the average value of the "time"-ordered product of $2n+4m$ field operators (cf. Equation (II.3)). All diagrams for $n+m \leq 2$ are shown in Fig. 5. Many of these diagrams are similar in their structure and give identical results when integrated. As an example we consider the following two diagrams, where (b)

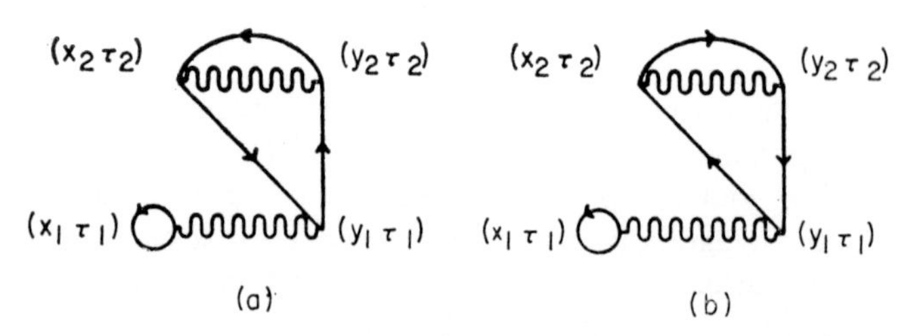

is obtained from (a) through the permutation of x_2 and y_2. Since the interaction potential $w(x, y)$ is symmetric as well as the associated product of

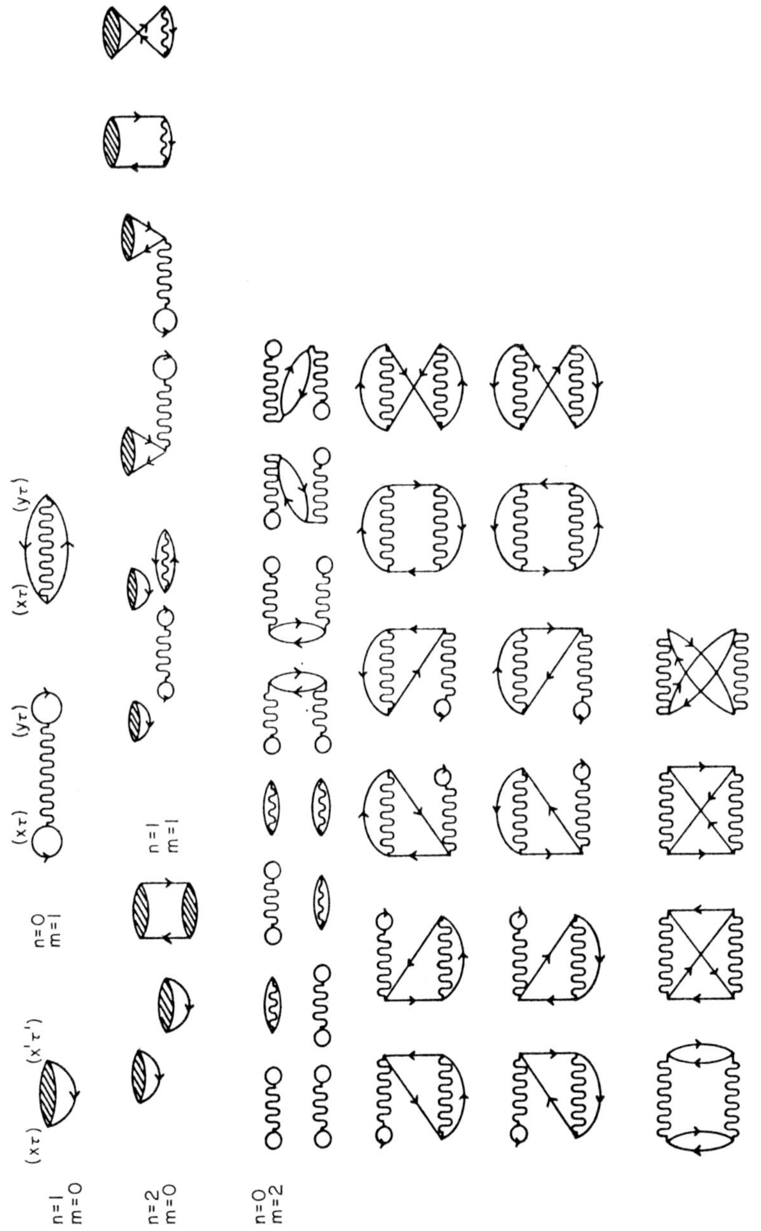

FIG. 5. All diagrams corresponding to the terms in the perturbation series in Equation 11–16 with $n+m \leq 2$.

contractions, the integral of the two diagrams will be the same. Similarly one can realize that the diagram,

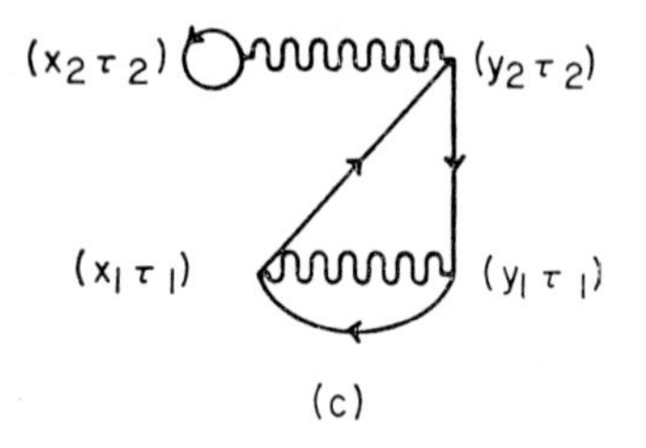

(c)

gives the same contribution as (a).

Diagrams that give the same result as a particular one have the same topology but differ in the specific labelling of points. To each diagram of a certain topological structure there is associated an integer which gives the number of distinct ways to label the points. Generally we expect that a factor of 2 arises from each interaction line $w(x, y)$ which cancels the factor $\frac{1}{2}$ in the definition of V_2. There will also be a factor $n!m!$ due to all possible assignments of "time"-variables. This will not be the case, however, when the diagram has some symmetry. We have for instance only one way of labelling the diagram:

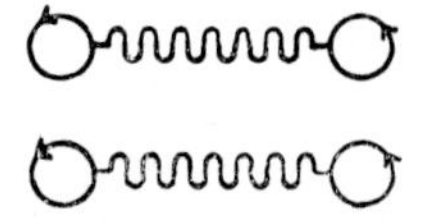

The symmetry of a diagram will determine the number of distinct ways it can be labelled and we will generally have to consider this as a special problem. The rules for evaluating contributions from topologically different unlabelled diagrams are given in Table 12. The concept of a fermion loop is important. It arises from the fact that contraction lines can be followed through a diagram and will form closed paths if we regard the one-particle potential $v(x', x)$ as part thereof. The sequence of contractions in such a circuit corresponds to a factor -1 from the determinantal form of contractions. A circuit

TABLE 12. Rules for Evaluation of Unlabelled Topologically Different Diagrams

1. A factor $v(x', x)$, $w(x, y)$, or $g(x\tau, x'\tau')$ corresponding to each element of the diagram.
2. A factor $(-)^{m+l}$, where m is the number of w-factors, and l is the number of fermion loops.
3. A factor $(1/r)$ where r is the number of ways one can label the diagram without changing its topology.
4. Integrate over all space and "time"-variables.

with only one contraction is also considered a fermion loop, which gives an additional factor $(-)^n$ in Equation (11.17).

As an example we evaluate the contribution from 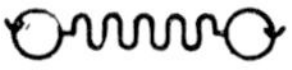 which is $-\frac{1}{2}\int dx\, dy\, d\tau\, g(x\tau, x\tau)\, w(x, y)\, g(y\tau, y\tau)$.

According to our definition of the "time"-ordered product at equal "times" we have that

$$g(x\tau, x\tau) = \langle\psi^\dagger(x)\psi(x)\rangle_0, \tag{11.18'}$$

and the integral is then

$$-\tfrac{1}{2}\beta\int dx\, dy\, \langle\psi^\dagger(x)\psi(x)\rangle_0\, w(x, y)\, \langle\psi^\dagger(y)\psi(y)\rangle_0,$$

which is the Coulomb self-interaction of the unperturbed charge density multiplied with $-\beta$.

Direct calculation of the partition function according to the prescription in this section is not a practical proposition, but the rules will now be used to derive perturbation expansions for other quantities.

Free Energy

The free energy is defined as

$$F = -\beta^{-1}\log Z(\beta), \tag{11.19}$$

or implicitly

$$Z(\beta) = \exp[-\beta F]. \tag{11.20}$$

We denote by F_0 the free energy of the unperturbed system with the partition function $Z_0(\beta)$ and obtain from Equation (11.5) that

$$\langle S(\beta)\rangle_0 = \exp[-\beta(F - F_0)]. \tag{11.21}$$

Obtaining the expansion of $(F - F_0)$ in a perturbation series is now very similar to a transformation from a moment expansion to a cumulant expansion.

The first order approximation to $\langle S(\beta)\rangle_0$ is from Equation (11.8)

$$\langle S(\beta)\rangle_0 = 1 - \beta\langle V\rangle_0 \tag{11.22}$$

and the first order contribution to the free energy is thus by comparison

$$F - F_0 = \langle V\rangle_0. \tag{11.23}$$

This first order term will, in an expansion of the exponential of Equation (11.21), occur as if it arose from an integral

$$[n!]^{-1}(-\beta\langle V\rangle_0)^n = (-)^n[n!]^{-1}\int_0^\beta\cdots\int_0^\beta d\tau_1\ldots d\tau_n\, \langle V(\tau_1)\rangle_0\ldots\langle V(\tau_n)\rangle_0. \tag{11.24}$$

Such a term can be identified on the left hand side of Equation (11.21), when the diagrammatic analysis has been carried out, as the contributions from diagrams with contractions at equal "times" only. These are disconnected diagrams in the sense that there are parts of them which can be integrated independently of the rest which means topologically that there can be no interaction or contraction lines joining them.

The factoring property of disconnected diagrams allows us to conclude that the free energy is given by *connected* diagrams only,

$$F = F_0 - \beta^{-1} \langle S(\beta) \rangle_c, \tag{11.25}$$

where the subscript c indicates the limitation to connected diagrams.

Electron Propagator

Generation of all connected topologically distinct diagrams is an unwieldy problem but it can be rationalized through the introduction of some auxiliary concepts. A simple diagram can be the source of many new ones where a simple element like a contraction is replaced by another part which begins and ends with a contraction and which is connected to other elements, for instance

On the other hand we can *define* the propagator $G(x\tau, x'\tau')$ as the sum of all connected diagrams, which begin with a particle (contraction) line at $(x'\tau')$ and end the particle line at $(x\tau)$ and where no other open ends occur. The perturbations under consideration here conserve the number of particles and it will thus be possible to follow the original particle line through a propagator diagram to the end.

Previously the electron propagator was defined in Equation (5.15) and the use of the same term in this context requires justification. This is readily done through the Fourier representation. Let us first consider the contraction, $g(x\tau, x'\tau')$, as a function of the relative argument $\tau - \tau'$ on the interval $(-\beta, \beta)$ and we find

$$g(x\tau, x'\tau') = \sum_n g_n(x, x') \exp\left[in\,\pi\,(\tau - \tau')/\beta\right] \tag{11.26}$$

with

$$\begin{aligned} g_{2m}(x, x') &= \tfrac{1}{2}\delta_{m0} \textstyle\sum_k u_k(x) u_k^*(x') \delta\,(\epsilon_k - \mu) \operatorname{tgh} \tfrac{1}{2}\beta\,(\epsilon_k - \mu) \\ g_{2m+1}(x, x') &= -\textstyle\sum_k u_k\,(x) u_k^*(x') / [\,(\epsilon_k - \mu)\beta + (2m+1)\pi\, i]. \end{aligned} \tag{11.27}$$

The limit when β tends to infinity, the zero temperature limit, of Equation (11.26) is

$$g(x\tau, x'\tau') = \sum_k u_k(x) u_k^*(x') (2\pi)^{-1} \int_{-\infty}^{\infty} dz\, e^{-iz(\tau-\tau')}/[iz+\mu-\epsilon_k]. \quad (11.28)$$

There may occur a singularity at $z=0$ since μ may tend to one of the ϵ_k's at the limit as discussed in Chapter 8. The result

$$\beta(\epsilon_k-\mu) = \log\,[1-\langle a_k^\dagger a_k\rangle_0] - \log\,\langle a_k^\dagger a_k\rangle_0, \quad (11.29)$$

follows from Equation (II.2) and can be used for determining limiting values. Similarly one can find from Equation (11.26) that the Fourier series on the right hand side becomes, for $\tau=\tau'$,

$$\begin{aligned} \sum_n g_n(x, x') &= -\sum_k u_k^*(x) u_k^*(x') \tfrac{1}{2}\mathrm{tgh}\, \tfrac{1}{2}\beta\,(\epsilon_k-\mu) \\ &= \tfrac{1}{2}\,\langle \psi^\dagger(x')\psi(x)\rangle_0 - \tfrac{1}{2}\,\langle \psi(x)\psi^\dagger(x')\rangle_0, \end{aligned} \quad (11.30)$$

in agreement with the property of a Fourier series to yield the arithmetic mean of left and right function values at a discontinuity.

The Fourier integral (11.28) is the inverse Laplace transform of the function

$$g(x, x'; \epsilon) = \sum_k u_k(x) u_k^*(x')/[\epsilon-\epsilon_k], \quad (11.31)$$

according to the formula

$$g(x\tau, x'\tau') = (2\pi)^{-1} \int_{\mu-i\infty}^{\mu+i\infty} d\epsilon\, g(x, x'; \epsilon) \exp\,[(\mu-\epsilon)(\tau-\tau')]. \quad (11.32)$$

A non-interacting system of electrons has an electron propagator with an energy representation as in Equation (11.31). It is natural to assume that the same relation will hold for the propagator $G(x\tau, x'\tau')$ and the previously defined one for an interacting system, Equation (5.15). Baym and Mermin have shown how to perform the necessary analytic continuations to ensure this result. The results have also been discussed by Luttinger and Ward and by Parry and Turner.

Our previous discussions of the electron propagator were concerned with its equation of motion. The diagrammatic approach leads to an equivalent integral equation. In order to reach this result we will introduce the concept of a self-energy diagram. When the initial and final contraction lines are removed from a propagator diagram, the remaining part is a *self-energy diagram*. Any self-energy diagram which cannot become disconnected through the removal of a contraction line is called a *proper self-energy diagram*. Such a diagram corresponds to a function of the arguments $(x'\,\tau')$ where the particle line enters and the arguments $(x\tau)$ where it exits. We define

$$M(x\tau, x'\,\tau') = \text{sum of all proper self-energy diagrams with particle line entering at } (x'\,\tau') \text{ and exiting at } (x\tau). \quad (11.33)$$

The lowest order proper self-energy diagrams are, $(x\tau)$ $(x'\tau')$

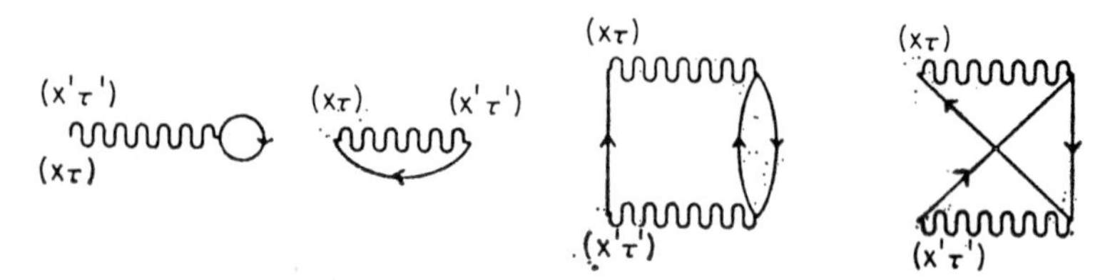

Any propagator diagram can be drawn as a sequence of contractions and proper self-energy parts,

= + M + M M + . . .

where a double line represents $G(x\tau, x'\tau')$. We write the summation of these terms in the form, where (1) stands for $(x_1\tau_1)$,

$$
\begin{aligned}
G(1, 1') = g(1, 1') + \int g(1, 2)d(2)M(2, 2')d(2') & [g(2', 1') \\
+ \int g(2', 3)d(3)M(3, 3')d(3') & [g(3', 1') \\
+ \ldots & \qquad (11.34) \\
= g(1, 1') + \int g(1, 2)d(2)M(2, 2')d(2')G(2', 1'), &
\end{aligned}
$$

which is the integral equation for the propagator in terms of the proper self-energy and the unperturbed propagator. The compact form of Equation (11.34) changes the problem of calculating $G(1, 1')$ into the problem of calculating $M(1, 1')$, the self-energy operator. One can ascertain through a tedious but straight forward procedure that both these functions depend on the relative argument $\tau - \tau'$ and can be analysed in Fourier series of the type (11.26). The integrations over intermediate "times" can then be carried out explicitly leading to an integral equation in space with an energy parameter. The relation (11.34) is known as the Dyson equation after a similar formula in quantum electrodynamics.

Polarization Propagator

Electron interaction lines will in this section be treated analogously to the way particle lines were dealt with in the previous section. A formal summation will again lead to an integral equation.

We consider all connected diagrams where particle lines have been deleted so that each diagram begins and ends with an interaction line and call the sum of these *the modified interaction*. Some low order diagrams of this type are the following,

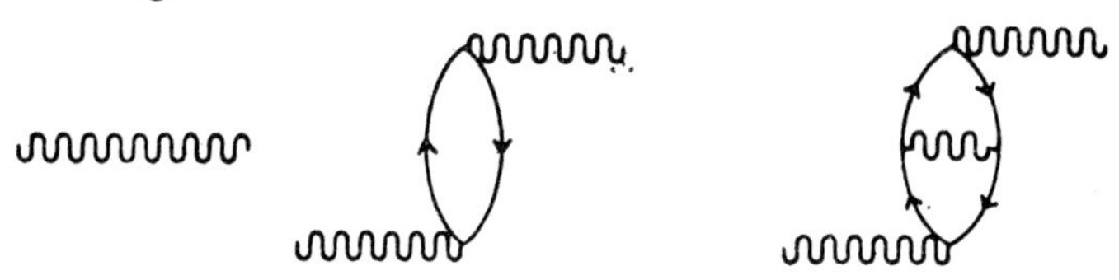

The part of any such diagram between the first and last interaction line is called a *polarization part* and if it does not become disconnected when any one of its interaction lines is removed it is a *proper polarization part*. The sum of all parts of the former or latter kind are named the polarization propagator and the proper or irreducible polarization propagator, respectively. We define thus

$P(x\tau, x'\tau')=$ the sum of all proper polarization parts where particle lines are entering and leaving at the points $(x\tau)$ and $(x'\tau')$.

All contributions to the modified electron interaction can now be written as a sequence of interaction lines and irreducible polarization propagator symbols. The formal summation of this series is expressed through the integral equation

$$\tilde{w}(1, 2)=w(1, 2)+\int w(1, 3)d(3)P(3, 4)d(4)\tilde{w}(4, 2), \tag{11.35}$$

where again we use a compact notation for space–"time" points and introduce the formula,

$$w(x\tau, x'\tau')=w(x, x')\,\delta(\tau-\tau'). \tag{11.36}$$

The polarization propagator $Q(1, 2)$ is similarly derived from the equation

$$Q(1, 2)=P(1, 2)+\int P(1, 3)w(3, 4)Q(4, 2)d(3)d(4). \tag{11.37}$$

These equations can also be handled by the Fourier technique and the functions obtained can be analytically continued to be related with previously defined Green's functions, in particular the particle-hole propagator of Chapter 5.

A more explicit account of the connections to polarization properties of the system can be obtained from considering as a term of the perturbation V_1 a local potential

$$\Phi=\int dx\, r(x)\psi^{\dagger}(x)\psi(x). \tag{11.38}$$

The diagrammatic expansion of the free energy can be classified according to the order in Φ and direct inspection of all diagrams leads to the following expression for terms through second order

$$\delta F = F + \int dx\, r(x) \langle \psi^{\dagger}(x)\psi(x)\rangle + \iint dx\, dx' r(x) r(x') \int d(\tau-\tau')\, Q(x\tau, x'\tau')/2\beta \tag{11.39}$$

which demonstrates Q's role of response function.

Renormalized Diagrams

The previous discussion led to the definition of several concepts and we will conclude this chapter with a few remarks on the calculation of the free energy. It was possible to develop integral equations for the electron propagator and the modified electron interaction, which express the sum of certain classes of diagrams. These sums can now be introduced into the perturbation series for the free energy. Propagator lines are introduced instead of contraction lines and modified electron interaction lines are put in place of the original electron interaction lines. This process is called renormalization and leads to a simplified form of the perturbation series. A renormalized diagram must not have any self-energy parts or polarization parts. Some of the simplest diagrams are

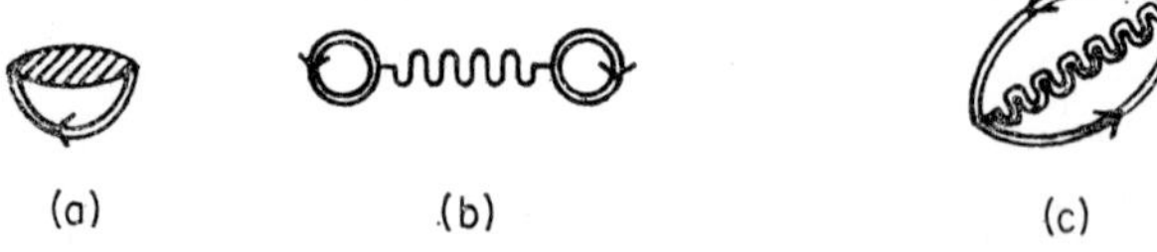

It is important to notice here that the interaction line in (b) is the bare electron interaction. This contribution to the energy represents the sum of all energy diagrams which become disconnected when one interaction line is removed. The propagators at each end of the interaction line represents the charge distribution in the system just as in Equation (11.18) we had the unperturbed charge distribution. The diagram (c) contains the modified interaction $\tilde{w}(x\tau, x'\tau')$.

At this point we are ready for the last piece of formal symbolism. It was stated above that the diagram (b) was the sum of all connected diagrams which become disconnected when an interaction line is removed. All remaining diagrams have the property that if the propagator lines and the modified interaction line, which are joined at a point of the graph, were removed the part that is left cannot be broken into disconnected parts by removal of an interaction line or a propagator line. These parts are called *irreducible vertex parts*, and their sum is the *irreducible vertex function*. Diagrams for this function connect three points, one where a particle line enters $(x'\tau')$, another where the particle line exits $(x\tau)$, and a third where an interaction line can be attached $(y\tau'')$. At the point $(y\tau'')$ there will be two propagator lines joined. We have for example

with the corresponding analytical form

$$\Gamma(x\tau, x'\tau'; y\tau'') = \delta(x-x')\,\delta(x-y)\,\delta(\tau-\tau')\,\delta(\tau-\tau'')$$
$$+\tilde{w}(x\tau, x'\tau')\,G(x\tau, y\tau'')\,G(y\tau'', x'\tau') + \ldots . \quad (11.40)$$

The irreducible vertex function is considerably more complicated than the previously discussed functions due to its dependence on three space-"time" points.

The final form of the free energy is now a closed expression where the difficulties of an infinite series have been eliminated, but where the implicit relations of the entering functions are now complications. The only diagrams that need to be considered are

which allow us to write the formula for the free energy as

$$F = F_0 + \beta^{-1}\Big[\int dx\,dx'\,d\tau v(x', x)\,G(x\tau, x'\tau)$$
$$+\tfrac{1}{2}\int dx\,dy\,d\tau G(x\tau, x\tau)\,w(x, y)\,G(y\tau, y\tau) \quad (11.41)$$
$$-\tfrac{1}{2}\int d(1)\,d(2)\,d(3)\,d(4)\,\tilde{w}(1, 4)\,G(1, 3)\,\Gamma(3, 2; 4)\,G(2, 1)\Big].$$

The factor $\frac{1}{2}$ and $-\frac{1}{2}$ that occur in the next to last and last term respectively of the right hand side of Equation (11.41) follow from the symmetry of the corresponding diagrams and the fact that the last term has one fermion loop.

The irreducible vertex function can also be used to define the self-energy operator. We find that

$$M(x'\tau', x\tau) = \delta(\tau'-\tau)v(x', x) + \delta(\tau'-\tau)\,\delta(x'-x)\int dy w(x, y)\,G(y\tau, y\tau)$$
$$-\int d(1)\,d(2)\,\tilde{w}(x\tau, 1)\,G(2, x\tau)\,\Gamma(x'\tau', 2; 1). \quad (11.42)$$

Similarly we can convince ourselves that the irreducible polarization propagator can be expressed as

$$P(1,2)=\int d(3)\,d(4)\,G(1,3)\,G(4,1)\,\Gamma(3,4;2). \qquad (11.43)$$

These relations and the previously obtained integral equations define a set of interlocked problems to which a solution can be found only by an iterative procedure which in most cases will be formidable.

The advantage of the formalism developed in this chapter lies mainly in the possibility of writing down in an efficient way a perturbation expansion for quantities of interest and to have a graphical structure of each term which may lead to an understanding of the relative importance of particular contributions. The functional relations between the propagators, the modified electronic interaction and the irreducible vertex function can also be derived by the technique of functional derivatives (cf. Martin and Schwinger, Kadanoff and Baym). It will be advantageous in the following to be able to use the diagrammatic method in conjunction with algebraic and functional methods to discuss various properties of electronic systems. The rules for evaluating a particular contribution will not be very important and we will consider diagrams in real time variables as well as in an energy representation. Mattuck has a rather complete discussion of the diagrammatic method in different representations.

Problem

Evaluate the contribution from the diagram 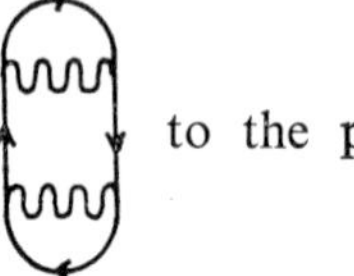 to the perturbation expansion in Equation (11.16) using the rules of Table 12.

Notes and Bibliography

Diagrammatic analysis was introduced by R. P. Feynman (1949) in *Phys. Rev.* **76**, 769, for time-dependent problems and by T. Matsubara (1955) in *Progr. Theoret. Phys.* (*Kyoto*) **14**, 351, for the temperature case. Wick's theorem occurs first in a paper by G. C. Wick (1950) in *Phys. Rev.* **80**, 268. The work by G. Baym and N. Mermin (1961) in *J. Math. Phys.* **2**, 232, J. M. Luttinger and J. C. Ward (1960) in *Phys. Rev.* **118**, 1417, and W. Parry and R. Turner (1964) in *Reports on Progres- in Physics*, **27**, 23, is relevant to the problem of analytic properties of propagators. Relations between propagators, modified electron interaction and the irreducible vertex function have been derived by P. C. Martin and J. Schwinger (1959) in *Phys. Rev.* **115**, 1342, and by L. P. Kadanoff and G. Baym (1962) in "Quantum Statistical Mechanics" (W. A. Benjamin Inc., New York), using the technique of functional derivatives. The simple idea of a cumulant expansion in mathematical statistics has been given a generalized treatment with applications to physics by R. Kubo (1962) in *J. Phys. Soc. Japan* **17**, 1100.

R. D. Mattuck (1967) has given an amusing and witty treatise of diagrammatic methods and the many-particle problem with the title "A Guide to Feynman Diagrams in the Many-body Problem" (McGraw-Hill Book Co., New York).

APPENDIX II

Wick's Theorem

We demonstrate here that the average value of a "time-ordered" product can be evaluated as the determinant of a set of so-called contractions. A contraction is defined as the average value of the "time-ordered" product of an annihilation and a creation operator in the interaction representation,

$$\begin{aligned} g(x\tau, x'\tau') &= -\langle T[\psi(x\tau)\psi^{\dagger}(x'\tau')]\rangle_0 \\ &= \sum u_k(x)\overset{*}{u_k}(x') \exp[-(\tau-\tau')(\epsilon_k-\mu)] \\ &\quad \times [\theta(\tau'-\tau)\langle a_k^{\dagger}a_k\rangle_0 - \theta(\tau-\tau')\langle a_k a_k^{\dagger}\rangle_0], \end{aligned} \tag{II.1}$$

$$\langle a_k^{\dagger}a_k\rangle_0 = 1-\langle a_k a_k^{\dagger}\rangle_0 = [1+\exp[\beta(\epsilon_k-\mu)]]^{-1}, \tag{II.2}$$

where we have the same notations as in Chapter 11.

The notation $\psi(j)=\psi(x_j\tau_j)$ will be used in the following in order to simplify the writings. By means of induction we will show that

$$(-)^n \langle T[\psi(1)\psi(2)\ldots\psi(n)\psi^{\dagger}(n')\ldots\psi^{\dagger}(2')\psi^{\dagger}(1')]\rangle_0 = \begin{vmatrix} g(1,1')\, g(1,2')\ldots g(1,n') \\ g(2,1')\, g(2,2')\ldots g(2,n') \\ \ldots \\ \ldots \\ g(n,1')\, g(n,2')\ldots g(n,n') \end{vmatrix}. \tag{II.3}$$

The proof is based on the following relations,

$$\langle Xa_k\rangle_0 = \langle a_k^{\dagger}a_k\rangle_0 \langle [X, a_k]_+\rangle_0, \tag{II.4}$$

$$\langle Xa_k^{\dagger}\rangle_0 = \langle a_k a_k^{\dagger}\rangle_0 \langle [X, a_k^{\dagger}]_+\rangle_0, \tag{II.5}$$

for a general operator X. These relations are derived from the transformation properties

$$a_k(\beta) = \exp[\beta(H_0-\mu N_{\text{op}})]\, a_k \exp[-\beta(H_0-\mu N_{\text{op}})] = a_k \exp[-\beta(\epsilon_k-\mu)], \tag{II.6}$$

$$a_k^{\dagger}(\beta) = \exp[\beta(H_0-\mu N_{\text{op}})]\, a_k^{\dagger} \exp[-\beta(H_0-\mu N_{\text{op}})] = a_k^{\dagger} \exp[\beta(\epsilon_k-\mu)], \tag{II.7}$$

and the invariance properties of the trace:

$$\begin{aligned}\langle Xa_k\rangle_0 &= Z_0^{-1}\, Tr\, Xa_k \exp\left[-\beta\,(H_0-\mu N_{\rm op})\right]\\ &= Z_0^{-1}\, Tr\, X \exp\left[-\beta\,(H_0-\mu N_{\rm op})\right] a_k(\beta)\\ &= \langle a_k(\beta)X\rangle_0 = \langle a_k X\rangle_0 \exp\left[-\beta\,(\epsilon_k-\mu)\right].\end{aligned} \tag{II.8}$$

Two different situations are encountered in the evaluation of the left hand side of Equation (II.3);

(i) an unprimed variable τ_j is less than all other "time"-variables,
(ii) a primed variable τ_j' is less than all other "time"-variables.

We consider those separately.

(i) $\tau_j = \min_k (\tau_k, \tau_k')$,

$$\begin{aligned}&(-)^n \langle T[\psi(1)\ldots\psi^\dagger(1')]\rangle_0\\ &= (-)^{n-j} \langle T[\psi(1)\ldots\psi(j-1)\psi(j+1)\ldots\psi^\dagger(1')]\,\psi(j)\rangle_0\\ &= \Sigma_k u_k(x_j) \exp\left[-\tau_j(\epsilon_k-\mu)\right] (-)^{n-j}\\ &\qquad\qquad \times \langle T[\psi(1)\ldots\psi(j-1)\psi(j+1)\ldots\psi^\dagger(1')]\,a_k\rangle_0\\ &= \Sigma_k u_k(x_j) \exp\left[-\tau_j(\epsilon_k-\mu)\right] \Sigma_\nu\, (-)^{n-j+\nu-1}\, [\psi^\dagger(\nu'), a_k]_+ \langle a_k^\dagger a_k\rangle_0\\ &\qquad \times \langle T[\psi(1)\ldots\psi(j-1)\psi(j+1)\ldots\psi^\dagger([\nu+1]')\psi^\dagger([\nu-1]')\ldots\psi^\dagger(1')]\rangle_0\\ &= \Sigma_\nu\, (-)^{\nu-j} \langle\psi^\dagger(\nu')\psi(j)\rangle_0\, (-)^{n-1} \langle T[\psi(1)\ldots\psi^\dagger(1')]\rangle_0\\ &= \Sigma_\nu\, (-)^{\nu+j}\, g(j,\nu')\, (-)^{n-1} \langle T[\psi(1)\ldots\psi^\dagger(1')]\rangle_0.\end{aligned} \tag{II.9}$$

It should be noticed that in the last two lines of Equation (II.9) the operators $\psi(j)$ and $\psi^\dagger(\nu')$ are not contained in the expectation value of the "time-ordered" product and that the replacement of the expectation value with the contraction is permitted since $\tau_j < \tau_\nu'$.

(ii) $\tau_j' = \min_k (\tau_k, \tau_k')$,

$$\begin{aligned}&(-)^n \langle T[\psi(1)\ldots\psi^\dagger(1')]\rangle_0\\ &= (-)^{n+j-1} \langle T[\psi(1)\ldots\psi^\dagger([j+1]')\psi^\dagger([j-1]')\ldots\psi^\dagger(1')]\,\psi^\dagger(j')\rangle_0\\ &= \Sigma_\nu\, (-)^{j+\nu-1} \langle\psi(\nu)\psi^\dagger(j')\rangle_0\, (-)^{n-1} \langle T[\psi(1)\ldots\psi^\dagger(1')]\rangle_0\\ &= \Sigma_\nu\, (-)^{j+\nu}\, g(\nu,j')\, (-)^{n-1} \langle T[\psi(1)\ldots\psi^\dagger(1')]\rangle_0.\end{aligned} \tag{II.10}$$

It follows from above that if the determinantal form (II.3) is valid when $n=p$ then it holds also for $n=p+1$ since Equations (II.9) and (II.10) are the expansions of the determinant of Equation (II.3) along a row and a column respectively, in terms of the elements and their minors. It is obvious that the formula is correct for $n=2$ and thus the induction proof is complete.

CHAPTER 12

Description of Some Processes Involving Photons

The general theory of the interaction between the electromagnetic radiation field and matter is a question of quantum electrodynamics, which subject is beyond the scope of this text. We want to set forth here a working description of electromagnetic processes in matter on the basis of propagator theory. This allows us to discuss several photon scattering and absorption problems on a common footing.

Quantization of the Radiation Field

A quantized formulation for the electromagnetic radiation field can be obtained similarly as for the electron field $\psi(\xi)$ in Chapter 4. The basic quantity to be considered is the vector potential $\bar{A}(\bar{r})$ which always will be chosen divergence free:

$$\text{div}\,\bar{A}=\nabla.\bar{A}=0. \tag{12.1}$$

Our development is thus specialized to the Coulomb gauge, a natural choice when the Coulomb interaction between particles is included in the hamiltonian of matter. The vector potential will be expanded in a complete set of plane waves, which are normalized over a volume V, such that

$$\bar{A}(\bar{r})=V^{-\frac{1}{2}}\sum_k \bar{A}(\bar{k})\exp(i\bar{k}.\bar{r}). \tag{12.2}$$

It follows from Equation (12.1) that the coefficients $\bar{A}(\bar{k})$ are perpendicular to the wave vector $\bar{k}$,

$$\bar{k}.\bar{A}(\bar{k})=0, \tag{12.3}$$

and can be expressed with the aid of two orthonormal vectors $\bar{n}_1(\bar{k})$ and $\bar{n}_2(\bar{k})$, which are orthogonal to $\bar{k}$. A convenient choice in terms of the polar angles of $\bar{k}$ is the following:

$$\begin{aligned}\bar{k}&=k\,(\sin\theta\cos\phi,\ \sin\theta\sin\phi,\ \cos\theta),\\ \bar{n}_1(\bar{k})&=(\cos\theta\cos\phi,\ \cos\theta\sin\phi,\ -\sin\theta),\\ \bar{n}_2(\bar{k})&=(\ -\sin\phi\ ,\quad \cos\phi\ ,\quad 0\ ),\end{aligned} \tag{12.4}$$

such that $\bar{n}_1(\bar{k})$, $\bar{n}_2(\bar{k})$, and $\bar{k}$ form a right-handed system of axes.

We define the amplitudes of the different components as

$$A_\lambda(\bar{k}) = \bar{n}_\lambda(\bar{k}) . \bar{A}(\bar{k}) \tag{12.5}$$

and will presently consider these as the dynamical variables.

The vector potential serves to define the electric and magnetic field operators as

$$\bar{E}(\bar{r}) = -c^{-1}\partial\bar{A}(\bar{r})/\partial t = -c^{-1}V^{-\frac{1}{2}} \sum_{\lambda\bar{k}} \dot{A}_\lambda (\bar{k})\bar{n}_\lambda(\bar{k}) \exp(i\bar{k} . \bar{r}), \tag{12.6}$$

and

$$\bar{B}(\bar{r}) = \text{curl } \bar{A}(\bar{r}) = -iV^{-\frac{1}{2}} \sum_{\lambda\bar{k}} (-)^\lambda k A_\lambda(\bar{k})\bar{n}_{\lambda\pm1}(\bar{k}) \exp(i\bar{k} . \bar{r}), \tag{12.7}$$

respectively. We define the energy of the field as the quantity

$$\begin{aligned} H_{ph} &= (8\pi)^{-1} \int d\bar{r} \, [\bar{E}^\dagger . \bar{E} + \bar{B}^\dagger . \bar{B}] \\ &= (8\pi c^2)^{-1} \sum_{\lambda\bar{k}} [\dot{A}_\lambda^\dagger (\bar{k})\dot{A}_\lambda(\bar{k}) + (kc)^2 A_\lambda^\dagger(\bar{k})A_\lambda(\bar{k})], \end{aligned} \tag{12.8}$$

which is an obvious generalization of the classical field energy but where the adjoint operators are brought in. This is convenient but increases the number of variables by a factor of two. We can eliminate the extra degree of freedom if we realize that the electric and magnetic fields are observables describable by means of hermitian operators. Thus it holds that

$$A_\lambda^\dagger(\bar{k}) = (-)^{\lambda+1} A_\lambda(-\bar{k}). \tag{12.9}$$

The most convenient way of defining independent operators is to form combinations as

$$b_\lambda(\bar{k}) = [i\dot{A}_\lambda(\bar{k}) + kcA_\lambda (\bar{k})] \, (8\pi kc^3)^{-\frac{1}{2}}, \tag{12.10}$$

where a suitable normalization factor has been introduced. It follows from Equations (12.9) and (12.10) that

$$\begin{aligned} b_\lambda^\dagger (-\bar{k}) &= [-i\dot{A}_\lambda^\dagger (-\bar{k}) + kcA_\lambda^\dagger (-\bar{k})] \, (8\pi kc^3)^{-\frac{1}{2}} \\ &= (-)^{\lambda+1} \, [-i\dot{A}_\lambda(\bar{k}) + kcA_\lambda(\bar{k})] \, (8\pi kc^3)^{-\frac{1}{2}}, \end{aligned} \tag{12.11}$$

which leads to the energy expression

$$H_{\text{ph}} = \tfrac{1}{2} \sum_{\lambda\bar{k}} kc \, [b_\lambda^\dagger(\bar{k})b_\lambda (\bar{k}) + b_\lambda(-\bar{k})b_\lambda^\dagger (-\bar{k})]. \tag{12.12}$$

The quantization requires certain commutation rules for the operators $b_\lambda(\bar{k})$ and $b_\lambda^\dagger(\bar{k})$ and these will be inferred from the equation of motion. The choice of the vector potential guarantees the validity of three of Maxwell's equations in vacuum. The remaining equation,

$$c^{-1} \, \partial\bar{E}/\partial t = \text{curl } \bar{B}, \tag{12.13}$$

gives the equation of motion for the operator $b_\lambda(\bar{k})$ as

$$\begin{aligned} idb_\lambda(\bar{k})/dt &= [-\ddot{A}_\lambda (\bar{k}) + ikc\dot{A}_\lambda(\bar{k})] \, (8\pi kc^3)^{-\frac{1}{2}} \\ &= [(kc)^2 A_\lambda(\bar{k}) + ikc\dot{A}_\lambda(\bar{k})] \, (8\pi kc^3)^{-\frac{1}{2}} \\ &= kcb_\lambda (\bar{k}). \end{aligned} \tag{12.14}$$

Thus we obtain the result that when H_{ph} of Equation (12.12) is taken as the hamiltonian it should hold that

$$[b_\lambda(\bar{k}), H_{\text{ph}}] = kcb_\lambda(\bar{k}). \tag{12.15}$$

This is consistent with the following commutation rules:

$$\begin{aligned} &[b_\lambda(\bar{k}), b_{\lambda'}(\bar{k}')] = 0, \\ &[b_\lambda(\bar{k}), b^\dagger_{\lambda'}(\bar{k}')] = \delta_{\lambda\lambda'}\delta_{\bar{k}\bar{k}'}, \\ &[b^\dagger_\lambda(\bar{k}), b^\dagger_{\lambda'}(\bar{k}')] = 0. \end{aligned} \tag{12.16}$$

They should be compared to the rules for fermion operators given in Equation (4.1) and have an analogous interpretation. The operator $b^\dagger_\lambda(\bar{k})$ applied to the vacuum state generates a state where one photon with propagation vector $\bar{k}$ and polarization vector $\bar{n}_\lambda(\bar{k})$ is present. Photons follow Bose–Einstein statistics as a consequence of the commutation rules (12.16).

The vector potential will now be given by the expansion

$$\bar{A}(\bar{r}) = \Sigma_{\lambda\bar{k}}(2\pi c/kV)^{\frac{1}{2}}\,[b_\lambda(\bar{k}) - (-)^\lambda b^\dagger_\lambda(-\bar{k})]\,\bar{n}_\lambda(\bar{k})\exp(i\bar{k}.\bar{r}), \tag{12.17}$$

and its equation of motion is determined by the hamiltonian (12.12) which equivalently can be written as

$$H_{\text{ph}} = \Sigma_{\lambda\bar{k}}kc\,[b^\dagger_\lambda(\bar{k})b_\lambda(\bar{k}) + \tfrac{1}{2}]. \tag{12.18}$$

The last part is the infinite, unobservable, zero point energy corresponding to the vacuum fluctuations.

Interactions Between Photons and Matter

The coupling of the electromagnetic field and matter arises from interaction terms in the total hamiltonian. They should lead to Maxwell's equations in the presence of charges and currents. In the gauge we have chosen only currents are important for the motion of photons; thus we have the basic commutator relation

$$[b_\lambda(\bar{k}), H] = kcb_\lambda(\bar{k}) - (2\pi/kcV)^{\frac{1}{2}}j_\lambda(\bar{k}), \tag{12.19}$$

with

$$j_\lambda(\bar{k}) = \bar{n}_\lambda(\bar{k}).\int d\bar{r}\,\bar{j}(\bar{r})\exp(-i\bar{k}.\bar{r}). \tag{12.20}$$

The components of the current operator, $j_\lambda(\bar{k})$, are themselves defined in terms of the vector potential in the nonrelativistic case that is discussed here as is seen in Equation (4.7).

The propagation of photons through matter is governed by these equations of motion and the related ones for matter operators and we can now determine the photon propagator as the Green's function

$$D_{\lambda'\lambda}(\bar{k}', \bar{k}; E) = \langle\langle b_{\lambda'}(\bar{k}'); b^\dagger_\lambda(\bar{k})\rangle\rangle_E. \tag{12.21}$$

In order to derive the final expression for this propagator we wish to make use of the following identity, which can be proved from the spectral representation of Green's functions in Equation (5.12):

$$\langle\langle [A, H]; B \rangle\rangle_E = \langle\langle A; [H, B] \rangle\rangle_E. \tag{12.22}$$

We find that

$$(E - k'c) D_{\lambda'\lambda}(\bar{k}', \bar{k}; E) = \delta_{\lambda'\lambda}\delta_{\bar{k}'\bar{k}} - (2\pi/k'cV)^{\frac{1}{2}} \langle\langle j_{\lambda'}(\bar{k}'); b_\lambda^\dagger(\bar{k}) \rangle\rangle_E, \tag{12.23}$$

and by means of the identity (12.22).

$$(E - kc) \langle\langle j_{\lambda'}(\bar{k}'); b_\lambda^\dagger(\bar{k}) \rangle\rangle_E = \langle [j_{\lambda'}(\bar{k}'), b_\lambda^\dagger(\bar{k})] \rangle - (2\pi/kcV)^{\frac{1}{2}} \langle\langle j_{\lambda'}(\bar{k}'); j_\lambda^\dagger(\bar{k}) \rangle\rangle_E. \tag{12.24}$$

The first term on the right hand side of Equation (12.24) is obtained from Equation (4.7) as

$$\langle [j_{\lambda'}(\bar{k}'), b_\lambda^\dagger(\bar{k})] \rangle = -(2\pi e^2/m^2kcV)^{\frac{1}{2}} \, \bar{n}_{\lambda'}(\bar{k}').\bar{n}_\lambda(\bar{k})\tilde{q}(\bar{k}' - \bar{k}) \tag{12.25}$$

with

$$\tilde{q}(\bar{k}) = \left\langle \int d\bar{r}\, q(\bar{r}) \exp(-i\bar{k}.\bar{r}) \right\rangle, \tag{12.26}$$

being the Fourier transform of the charge density operator expectation value.

The second term on the right hand side of Equation (12.24) contains the current propagator. This Green's function will in the following be approximated with its counterpart derived from the equations of motion of matter. Thus we disregard the term in the current operator which contains the vector potential. This approximation eliminates from the description the possibility of inelastic processes since the matter state will be the same after the collision as it was before. The total absorption is, however, always possible to calculate from the forward scattering amplitude by means of the optical theorem as will be seen in the following section.

The approximation also eliminates multiple scattering processes and is thus appropriate only for a dilute gas of molecules where radiative interactions can be neglected. A macroscopic theory can then be derived in the ordinary way.

Extraction of the information on a photon scattering process requires an analysis of wave packets, where the initial and final states are represented as superpositions of states with wave and polarization vectors nearly identical. We will implicitly apply such reasoning and assume that a wave packet with wave vector $\bar{k}$ and polarization λ is emitted at time 0. It will take this packet a time L/c to traverse the system when the volume V has linear dimensions L.

The probability amplitude that the wave packet can be detected with polarization λ', wave vector $\bar{k}'$, and the same energy $k'c=kc$ is then obtained from the Fourier integral as

$$\begin{aligned}\langle b_{\lambda'}(\bar{k}', L/c)\, b_{\lambda}^{\dagger}(\bar{k}, 0)\rangle &= (i/2\pi)\int dE\, D_{\lambda'\lambda}(\bar{k}', \bar{k}; E)\exp(-iEL/c)\\ &= \exp(-ikL)\,[\delta_{\lambda'\lambda}\delta_{\bar{k}'\bar{k}} - (2\pi iL/kc^2V)\,[(e/m)\tilde{q}(\bar{k}'-\bar{k})\\ &\qquad \times \bar{n}_{\lambda'}(\bar{k}').\bar{n}_{\lambda}(\bar{k}) + \langle\langle j_{\lambda'}(\bar{k}'); j_{\lambda}^{\dagger}(\bar{k})\rangle\rangle_{kc}]]. \qquad (12.27)\end{aligned}$$

The exponential phase factor can be seen to be unimportant for the probability in this formulation and goes together with a similar phase factor from the plane waves in position space to give a sharply peaked function which guarantees that the photon wave packets travel a distance L in the time L/c.

Absorption of Light

The probability amplitude calculated in Equation (12.27) gives direct information on the absorption processes. We assume that the quantization volume V is such that only one molecule is contained therein. Equivalently we can take the specific volume per molecule as the quantization volume. The probability that a photon with wave vector $\bar{k}$ and polarization vector $\bar{n}_{\lambda}(\bar{k})$ can pass this volume without being scattered is [cf. Equation (12.27)]

$$|\langle b_{\lambda}(\bar{k}) b_{\lambda}^{\dagger}(\bar{k})\rangle|^2 = 1 + (4\pi L/kc^2V)\ \mathrm{Im}\ \langle\langle j_{\lambda}(\bar{k}); j_{\lambda}^{\dagger}(\bar{k})\rangle\rangle_{kc}, \qquad (12.28)$$

when quadratic terms in (L/V) are discarded. The absorption coefficient a is the decrease in intensity per unit length and we obtain the result

$$a = -(4\pi/kc^2V)\ \mathrm{Im}\ \langle\langle j_{\lambda}(\bar{k}); j_{\lambda}(\bar{k})\rangle\rangle_{kc}, \qquad (12.29)$$

for the general case. It should be observed that scattering also includes emission processes and that this expression for the absorbance holds only for the ground state of matter.

It is possible in many cases to assume that the wavelength of the radiation is much longer than the typical dimensions of the molecule such that the exponential factor in the integral (12.20) can be put equal to unity. We will further assume that only electrons are considered so that

$$\int d\bar{r}\bar{j}(\bar{r}) = e\bar{p}/m, \qquad (12.30)$$

where $\bar{p}$ is the linear momentum operator for the electrons. Thus the basic Green's functions to consider are the elements of the tensor $\langle\langle \bar{p}; \bar{p}\rangle\rangle_E$. This tensor can be related to the similar one involving the electron dipole moment operator defined as

$$\bar{R} = \int d\bar{r}\ \bar{r}q(\bar{r}), \qquad (12.31)$$

from Equation (4.6). It follows that

$$[\bar{R}, H] = ie\bar{p}/m, \tag{12.32}$$

and that consequently

$$E\langle\langle \bar{R}; \bar{R} \rangle\rangle_E = (ie/m) \langle\langle \bar{p}; \bar{R} \rangle\rangle_E, \tag{12.33}$$

as well as

$$E\langle\langle \bar{p}; \bar{R} \rangle\rangle_E = \langle [\bar{p}, \bar{R}] \rangle - (ie/m) \langle\langle \bar{p}; \bar{p} \rangle\rangle_E. \tag{12.34}$$

The first term on the right hand side of Equation (12.34) is purely imaginary and proportional to the unit tensor and the total charge in the system. The second term can thus be expressed in terms of the so-called polarizability tensor, $-\langle\langle \bar{R}; \bar{R} \rangle\rangle_E$, and a real term. The spherically symmetric component of the polarizability tensor,

$$\alpha(E) = -(1/3)\left[\langle\langle R_x; R_x \rangle\rangle_E + \langle\langle R_y; R_y \rangle\rangle_E + \langle\langle R_z; R_z \rangle\rangle_E\right], \tag{12.35}$$

is the frequency dependent polarizability for the molecule. The absorption coefficient can now be expressed in terms of $\alpha(E)$ when it is assumed that in a dilute gas of molecules there is random orientation with respect to the direction of the photon. Thus one obtains from Equations (12.29)–(12.35) that

$$a = (4\pi k/V) \operatorname{Im} \alpha(kc), \tag{12.36}$$

in agreement with the semi-classical treatment.

Optical Rotatory Dispersion and Circular Dichroism

The scattering formalism can conveniently be used also to describe the phenomena of optical rotatory dispersion and circular dichroism. A photon with propagation vector $\bar{k}$ is scattered from polarization $\bar{n}_1(\bar{k})$ to the polarization $\bar{n}_2(\bar{k})$ with the transition amplitude

$$\langle b_2(\bar{k}, L/c) b_1^\dagger(\bar{k}, 0) \rangle = \exp(-ikL)\left[-(2\pi i L/kc^2 V) \langle\langle j_2(\bar{k}); j_1^\dagger(\bar{k}) \rangle\rangle_{kc}\right]. \tag{12.37}$$

The photon wave in the forward direction will thus have the polarization vector

$$\bar{n}_{\text{final}} = \bar{n}_1(\bar{k}) - \bar{n}_2(\bar{k})\, \Phi L, \tag{12.38}$$

where

$$\Phi = (2\pi i/kc^2 V) \langle\langle j_2(\bar{k}); j_1^\dagger(\bar{k}) \rangle\rangle_{kc} \tag{12.39}$$

is called the complex rotatory power. It is assumed that there is little scattering or absorption so that the coefficient for $\bar{n}_1(\bar{k})$ can be taken as unity. The plane of polarization of a beam of photons will be turned the angle $\operatorname{Re}\Phi$ per unit length of a medium with V^{-1} molecules per unit volume. Such a beam will acquire an ellipticity equal to $-\operatorname{Im}\Phi$ per unit length under these conditions. The comparison of these expressions and those commonly used requires again a limiting procedure for small wave vectors $\bar{k}$. We have that

$$j_\lambda(\bar{k})=(e/m)\int d\xi\psi^\dagger(\xi)\,[1-ikr_k-\ldots]\,[-i\nabla_\lambda\psi(\xi)], \qquad (12.40)$$

where r_k is the component of $\bar{r}$ along $\bar{k}$ and ∇_λ similarly is the component along $\bar{n}_\lambda(\bar{k})$. The tensor operator arising from the products $r_k(-i\nabla_\lambda)$ can be separated into a spherically symmetric part, a part which transforms as an axial vector, and a part transforming as spherical harmonics of order 2. The axial vector part is the orbital angular momentum and from the relation between the propagation vector and the polarization vector we obtain that

$$\begin{aligned} r_k(-i\nabla_\lambda) &= \tfrac{1}{2}\,[r_k(-i\nabla_\lambda)-r_\lambda\,(-i\nabla_k)]+\tfrac{1}{2}\,[r_k(-i\nabla_\lambda)+r_\lambda(-i\nabla_k)] \\ &= \tfrac{1}{2}\,(-)^{\lambda\pm 1}l_{\lambda\pm 1}+T^{(2)}_{k\lambda}. \end{aligned} \qquad (12.41)$$

It follows from this that a product of matrix elements of $j_2(\bar{k})$ and $j_1^\dagger(\bar{k})$ averaged over molecular orientations, or equivalently over propagation directions, will be proportional to the scalar product of angular and linear momentum matrix elements to first order in the magnitude of the wave vector. The formula for the complex optical rotatory power is then

$$\Phi=-(\pi e^2/3m^2c^2V)\,\Sigma_\nu\,[\langle\langle L_\nu;p_\nu\rangle\rangle_{kc}+\langle\langle p_\nu;L_\nu\rangle\rangle_{kc}]. \qquad (12.42)$$

The total orbital angular momentum operator $\bar{L}$ also defines the magnetic dipole moment operator,

$$\bar{M}=(e/2mc)\bar{L}, \qquad (12.43)$$

and by use of similar formulae as Equation (12.33) we obtain the final expression as

$$\Phi=(2\pi ik/3V)\,\Sigma_\nu\,[\langle\langle R_\nu;\,M_\nu\rangle\rangle_{kc}-\langle\langle M_\nu;\,R_\nu\rangle\rangle_{kc}], \qquad (12.44)$$

which can be seen to be equivalent to the semi-classical derivation.

The derivation given above has to be modified when the molecules are not randomly oriented. A particularly important case of this kind occurs when a constant magnetic field is used to induce optical activity.

Spin Resonance Absorption

The previous discussion of absorption of photons derived from the part of the current operator that involves the velocity operators of the particles. This section will be devoted to a discussion of the effects arising from the spin operator contributions to the current. We obtain for instance from the expression (4.7) that

$$\begin{aligned} j_\lambda(\bar{k}) &\simeq (e/2m)\int d\xi\,\exp(-i\bar{k}.\bar{r})\,\bar{n}_\lambda(\bar{k}).\text{curl}\,[\psi^\dagger(\xi)\bar{\sigma}\psi(\xi)] \\ &= (iek/m)\,(-)^\lambda\bar{n}_{\lambda\pm 1}(\bar{k}).\bar{S} \end{aligned} \qquad (12.45)$$

where a partial integration and the orthogonality properties of the propagation and polarization vectors have been used and where the long wavelength limit is taken to obtain the spin operator $\bar{S}$. In the expression for the absorption coefficient we will obtain the components of the tensor Green's function $\langle\langle\bar{S}; \bar{S}\rangle\rangle_{kc}$ instead of the polarizability. Absorption will not occur unless a magnetic field is applied so that $\bar{S}$ is no longer a constant of the motion.

Nuclear spins can also contribute to the current operator for a molecule so that instead of the electron spin magnetic moment operator $(e\bar{S}/m)$ there occurs the sum of nuclear magnetic moment operators $\gamma_s\bar{I}_s$, given in terms of the nuclear spin operators $\bar{I}_s$ and the corresponding gyromagnetic ratios γ_s. The total current operator in these cases is then

$$j_\lambda(\bar{k}) \cong ik(-)^\lambda \bar{n}_{\lambda\pm1}(\bar{k}) . [(e\bar{S}/m) + \textstyle\sum_s \gamma_s \bar{I}_s]. \tag{12.46}$$

This current operator is proportional to the magnetic moment operator arising from spin operators and we expect that a more complete expression would involve also the orbital angular momentum operator from Equation (12.43). This is indeed also the case but some care must be exercised in the treatment of the associated tensor term $T^{(2)}$ of Equation (12.41) since it will contribute to the result. This is essentially a problem of gauge invariance.

It is often permissible to ignore the orbital contribution to the magnetic moment operator in spin resonance problems and hence we find that the absorption is governed by the tensor Green's function defined with the moment operators of Equation (12.46). This is the magnetic susceptibility tensor with the definition

$$\chi(kc) = \langle\langle\bar{M}^s; \bar{M}^s\rangle\rangle_{kc}. \tag{12.47}$$

The total spin magnetic moment operator is here denoted $\bar{M}^s$:

$$\bar{M}^s = (e\bar{S}/m) + \textstyle\sum_s \gamma_s \bar{I}_s. \tag{12.48}$$

A similar expression holds when the orbital moment is included but then a new term must be added in Equation (12.47). It arises from the tensor $T^{(2)}$ and the charge density term where a power series expansion of the exponential also should be used. This contribution will be considered in some detail in Chapter 13.

Problem

Show that the quadrupole moment operator for an electronic system,

$$\int d\bar{r}\, \bar{r}\bar{r}\, q(\bar{r}),$$

is related to the tensor $T^{(2)}$ of Equation (12.41) through a commutator with the hamiltonian.

Notes and Bibliography

The quantization of the electromagnetic field is the topic of numerous texts. We like to mention the one by the pioneer P. A. M. Dirac: "The Principles of Quantum Mechanics" (Oxford, 1958). Definitions of polarizability, optical rotatory power, and susceptibility are given by H. Eyring, J. Walter, and G. E. Kimball in "Quantum Chemistry" (J. Wiley and Sons, New York, 1944).

CHAPTER 13

Nuclear Displacements, Nuclear Spins and Magnetic Fields

The formalism developed in the previous chapters is employed here to describe the influence of various external fields on molecular electronic systems. We limit ourselves to treatments within second order perturbation theory in the external influences, and we assume the general form of the hamiltonian

$$H = H_0 + \lambda V_1 + \lambda^2 V_2, \tag{13.1}$$

where we take H_0 to be the hamiltonian for the electrons in the system in the absence of external perturbations but including the electron interaction terms. The perturbations V_1 and V_2 are the first and second order contributions to the hamiltonian of the external perturbation respectively. It should be observed that these definitions differ from those used in Chapter 11 but that the general expressions (11.7), (11.16), and (11.21) still are valid. Thus we have for the free energy that

$$F = F_0 + \lambda \langle V_1 \rangle_0 + \lambda^2 \left[\langle V_2 \rangle_0 + \tfrac{1}{2}\beta \langle V_1 \rangle^2 - \tfrac{1}{2}\beta^{-1} \int_0^\beta \int_0^\beta d\tau \, d\tau' \, \langle T[V_1(\tau) V_1(\tau')] \rangle_0 \right] \tag{13.2}$$

through second order in the perturbation parameter λ.

The last term on the right hand side of Equation (13.2) can be related to the Green's function $\langle\langle V_1; V_1 \rangle\rangle_E$ and we wish to express the result in terms thereof. It will only be necessary to consider perturbations where V_1 is a boson-type operator and we have then for the spectral resolution of the Green's function from Equation (5.12)

$$\langle\langle V_1; V_1 \rangle\rangle_E = \sum_{nm} |\langle n | V_1 | m \rangle|^2 \left[\frac{\rho_n}{E - E_m + E_n + i\eta} - \frac{\rho_m}{E - E_m + E_n - i\eta} \right]. \tag{13.3}$$

The integrand in Equation (13.2) is obtained from the definition (11.7) and the density operator

$$\langle n | \rho | m \rangle = \delta_{nm} \rho_n = \delta_{nm} \exp [\beta (F_0 - E_n)], \tag{13.4}$$

as

$$\langle T[V_1(\tau) V_1(\tau')] \rangle_0 = \sum_{nm} | \langle n | V_1 | m \rangle |^2 \, [\rho_n \theta(\tau - \tau') + \rho_m \theta(\tau' - \tau)] \times \exp [(\tau - \tau') (E_n - E_m)]. \tag{13.5}$$

It obtains for the free energy that

$$F=F_0+\lambda \langle V_1\rangle_0+\lambda^2 [\langle V_2\rangle_0+\tfrac{1}{2}\beta \langle V_1\rangle_0^2+\tfrac{1}{2}\mathrm{Re} \langle\langle V_1; V_1\rangle\rangle_{E=0}]. \qquad (13.6)$$

Special care needs to be exercised in the limit of zero temperature, that is when β tends to infinity, when the ground state of the system is degenerate, as in the case of a paramagnetic molecule. These problems will be dealt with when they are encountered in particular applications.

Application to Nuclear Displacements

This section contains a description of force constants in terms of Green's functions from the development above. It will be assumed that the electron-nuclear interaction in the hamiltonian is given as

$$-\sum_g eZ_g \int d\bar{r}\, q(\bar{r})/|\bar{r}-\bar{R}_g|$$

and that we relate the nuclear positions to equilibrium positions $\bar{R}_g^0$ in the following way:

$$\bar{R}_g=\bar{R}_g^0+\lambda\bar{u}_g. \qquad (13.7)$$

The relevant perturbation operators are obtained by a Taylor series expansion in terms of λ as

$$V_1=-\sum_g eZ_g \int d\bar{r}\, [(\bar{u}_g . \nabla)\, q(\bar{r})]/|\bar{r}-\bar{R}_g^0|, \qquad (13.8)$$

and

$$V_2=-\sum_g eZ_g \int d\bar{r}\, [(\bar{u}_g . \nabla)^2 q(\bar{r})]/|\bar{r}-\bar{R}_g^0|. \qquad (13.9)$$

An equilibrium nuclear configuration is determined by a vanishing first order contribution to the energy,

$$\langle V_1\rangle_0=0, \qquad (13.10)$$

and we obtain for the free energy for small displacements from the equilibrium

$$F=F_0+\tfrac{1}{2} \sum_{gh}\bar{u}_g . \bar{\bar{D}}_{gh} . \bar{u}_h, \qquad (13.11)$$

The tensor $\bar{\bar{D}}$ is defined as

$$\bar{\bar{D}}_{gh}=e^2 Z_g Z_h \int d\bar{r}\, d\bar{r}'\, |\bar{r}-\bar{R}_g^0|^{-1}\, \mathrm{Re}\langle\langle \nabla q(\bar{r}); \nabla q(\bar{r}')\rangle\rangle_{E=0} |\bar{r}'-\bar{R}_h^0|^{-1}$$
$$-eZ_g\delta_{gh} \int d\bar{r}\, |\bar{r}-\bar{R}_g^0|^{-1} \langle \nabla\nabla q(\bar{r})\rangle_0 \qquad (13.12)$$

where the auxiliary parameter λ has been put equal to unity.

Force constants can thus be calculated when the polarization propagator is known as was demonstrated in Equation (11.39). It should be observed that $q(\bar{r})$ is a tensor operator in spin space of rank zero. Thus only matrix elements of $q(\bar{r})$ between states of equal multiplicity contribute to the spectral representation of the propagator. Particularly it holds that for a singlet ground state at zero temperature only excited singlet states are included. This case leads to several simplifications in the approximation to the polarization propagator that can be derived from Equation (7.63).

It is assumed that a closed shell ground state Hartree–Fock solution is obtained with spin orbitals labelled $(k\frac{1}{2})$ and $(k-\frac{1}{2})$. The particle-hole operators $q_\nu^\dagger$ of Equation (7.55) will here be used in the following combinations,

$$q_\nu^\dagger(00)=(\tilde{a}_{k\frac{1}{2}}^\dagger\tilde{a}_{l\frac{1}{2}}+\tilde{a}_{k-\frac{1}{2}}^\dagger\tilde{a}_{l-\frac{1}{2}})/\sqrt{2}, \tag{13.13}$$

$$\begin{aligned} q_\nu^\dagger(11)&=-\tilde{a}_{k\frac{1}{2}}^\dagger\tilde{a}_{l-\frac{1}{2}},\\ q_\nu^\dagger(10)&=(\tilde{a}_{k\frac{1}{2}}^\dagger\tilde{a}_{l\frac{1}{2}}-\tilde{a}_{k-\frac{1}{2}}^\dagger\tilde{a}_{l-\frac{1}{2}})/\sqrt{2},\\ q_\nu^\dagger(1-1)&=\tilde{a}_{k-\frac{1}{2}}^\dagger\tilde{a}_{l\frac{1}{2}}, \end{aligned} \tag{13.14}$$

which are tensor operators of rank zero and one, respectively, in spin space. The Hartree–Fock orbitals are chosen to be real and we can express the perturbation V_1 in terms of these particle-hole operators as

$$V_1=\Sigma_\nu(V_1)_\nu[q_\nu(00)+q_\nu^\dagger(00)]+\text{other terms} \tag{13.15}$$

where the other terms that are indicated do not contribute to the second order energy in the approximation here.

According to Equation (13.8) we have that the matrix elements are

$$(V_1)_\nu=[\,[\tilde{a}_{k\frac{1}{2}},V_1],\tilde{a}_{l\frac{1}{2}}^\dagger]_+\sqrt{2}. \tag{13.16}$$

There will further be a separation of the matrices **B** and **C** defined in Equations (7.60) and (7.61) such that we find

$$\begin{aligned}\langle[[q_\nu(SM),H],q_{\nu'}^\dagger(S'M')]\rangle&={}^{S}B_{\nu\nu'}\delta_{SS'}\delta_{MM'}\\ &=\delta_{SS'}\delta_{MM'}[\delta_{\nu\nu'}(\epsilon_k-\epsilon_l)-(kk'|l'l)-2(S-1)\,(kl|l'k')], \end{aligned}\tag{13.17}$$

and

$$\begin{aligned}\langle[[q_\nu(SM),H],q_{\nu'}(S'-M')]\rangle&=-{}^{S}C_{\nu\nu'}\delta_{SS'}\delta_{MM'}\\ &=\delta_{SS'}\delta_{MM'}[(kl'|k'l)+2\,(S-1)\,(kl|k'l')].\end{aligned}\tag{13.18}$$

Each block in the propagator matrix (7.63) can now be inverted separately.

The special case for the polarization propagator at $E=0$ and the form assumed by V_1 in Equation (13.15) leads to the expression for the second order energy as

$$\tfrac{1}{2}\text{Re}\langle\langle V_1;V_1\rangle\rangle_{E=0}=-\Sigma_{\nu\nu'}\,(V_1)_\nu[{}^0\mathbf{B}+{}^0\mathbf{C}]_{\nu\nu'}^{-1}\,(V_1)_{\nu'}, \tag{13.19}$$

and it is clear that this expression is useful only if the Hartree–Fock state is stable as discussed in Chapter 7 (pp. 44–46).

Calculations such as those presented above are strictly applicable only to situations where the basis set for the representation of the field operator is complete or at least independent of the nuclear positions. Most molecular calculations are performed in limited basis sets such as atomic orbitals which strongly depend upon the nuclear positions. Then it can be expected that the exact expression (13.11) neither is well approximated with Equation (13.19) nor equals the direct Taylor series expansion of a ground state energy calculated for varying nuclear configurations.

External Magnetic Fields

Static magnetic fields will produce energy changes in an electronic system that can be handled directly in the formalism given in the beginning of this chapter when we consider the system at zero temperature and assume a nondegenerate ground state. The case of paramagnetic systems is not treated in this section. The perturbations V_1 and V_2 will now be expressed in terms of the vector potential $\bar{A}(\bar{r})$ for the external field, which normally is chosen such that

$$\operatorname{div} \bar{A}=0. \tag{13.20}$$

Then we obtain, with λ taken as $1/c$,

$$V_1=-\int d\bar{r}\, \bar{A}(\bar{r}).\bar{j}^0(\bar{r})-\int d\bar{r}\, \bar{B}(\bar{r}).\bar{m}(\bar{r}), \tag{13.21}$$

and

$$V_2=(e/2m)\int d\bar{r}\, \bar{A}^2(\bar{r})q(\bar{r}). \tag{13.22}$$

Here we have defined the current density $\bar{j}^0(\bar{r})$ from the momentum density, cf. Equation (4.7),

$$\bar{j}^0(\bar{r})=(e/2m)\sum_{\text{spin}}[(i\nabla\psi^\dagger(\xi))\,\psi(\xi)-i\psi^\dagger(\xi)\,\nabla\psi(\xi)]. \tag{13.23}$$

The magnetic moment density is obtained as

$$\bar{m}(\bar{r})=(e/2m)\sum_{\text{spin}}\psi^\dagger(\xi)\bar{\sigma}\psi(\xi), \tag{13.24}$$

and we get the last term of V_1 by means of a partial integration from the corresponding term in the general expression, $-\int \bar{A}.\bar{j}\,d\bar{r}$, for the derivative $\partial H/\partial\lambda$. Similarly we obtain V_2 from the term in this expression which is inversely proportional to c.

The operators of Equations (13.23) and (13.24) are tensor operators of rank zero and one, respectively, in spin space. Since we consider only nondegenerate ground states at the zero temperature limit it follows that the second order energy from $\operatorname{Re}\langle\langle V_1; V_1\rangle\rangle_{E=0}$ arises from two separate terms corresponding to the different ranks of these tensor operators. Thus we write

$$F=F_0+\tfrac{1}{2}c^{-2}\Big[2\langle V_2\rangle_0+\mathrm{Re}\int d\bar{r}\,d\bar{r}'\bar{A}(\bar{r}).\langle\langle \bar{j}^0(\bar{r});\bar{j}^0(\bar{r}')\rangle\rangle_{E=0}.\bar{A}(\bar{r}')$$
$$+\mathrm{Re}\int d\bar{r}\,d\bar{r}'\bar{B}(\bar{r}).\langle\langle \bar{m}(\bar{r});\bar{m}(\bar{r}')\rangle\rangle_{E=0}.\bar{B}(\bar{r}')\Big], \qquad (13.25)$$

which will form the basis for our further discussion.

It is instructive to check the gauge invariance of the result (13.25). A transformation of the vector potential such that

$$\bar{A}\rightarrow\bar{A}'+\nabla\phi, \qquad (13.26)$$

where $\phi=\phi(\bar{r})$ is a function of position coordinates only, leads to changes in V_1 and V_2. These are conveniently expressed in terms of an auxiliary operator

$$\Phi=\int d\bar{r}\,\phi(\bar{r})q(\bar{r}), \qquad (13.27)$$

which gives, when commuted with the hamiltonian H_0,

$$[\Phi, H_0]=i\int d\bar{r}\,\nabla\phi.\bar{j}^0(\bar{r}), \qquad (13.28)$$

and, when commuted with V_1,

$$[\Phi, V_1]=(-ie/m)\int d\bar{r}\,\bar{A}(\bar{r}).\nabla\phi q(\bar{r}). \qquad (13.29)$$

Moreover it holds that

$$[\Phi, [\Phi, H_0]\,]=-(e/m)\int d\bar{r}|\,\nabla\phi|^2q(\bar{r}). \qquad (13.30)$$

The changes in the operators V_1 and V_2 can now be written in the compact form

$$V_1\rightarrow V_1'+i[\Phi, H_0], \qquad (13.31)$$

and

$$V_2\rightarrow V_2'+i[\Phi, V_1]+\tfrac{1}{2}[\Phi, [\Phi, H_0]], \qquad (13.32)$$

which indicates that the gauge transformation for the states is generated by the operator Φ. It follows from Equation (5.8) that

$$\langle\langle[A, H_0]; B\rangle\rangle_{E=0}=-\langle[A, B]\rangle_0, \qquad (13.33)$$

which together with Equations (13.31) and (13.32) guarantees that Equation (13.25) gives the same result for $\bar{A}'$ as for $\bar{A}$.

It is a consequence of Equation (13.25) that

$$F\leq F_0+c^{-2}\langle V_2'\rangle_0, \qquad (13.34)$$

for any choice of vector potential $\bar{A}'(\bar{r})$. This has been used to seek a function $\phi(\bar{r})$ which minimizes the upper bound to F. It is possible to satisfy the equality in relation (13.34) only for special cases by this procedure.

The magnetic susceptibility for a constant homogeneous magnetic field $\bar{B}=(00B)$ is obtained as, with $\bar{A}=\frac{1}{2}\bar{B}\times\bar{r}$,

$$\chi=\mathrm{Re}\,\langle\langle M_z;\,M_z\rangle\rangle_{E=0}-(e^2/4mc^2)\int d\bar{r}(x^2+y^2)\,\langle q(\bar{r})\rangle_0, \qquad (13.35)$$

which might be averaged over orientations to yield the spherical component of the tensor. The magnetic moment operator $\bar{M}$ was defined in Equation (12.43) and for the case we are considering there will be no contributions from the last term of Equation (12.25) although they could be formally included in the first term by a redefinition of $\bar{M}$. The form (12.47) which is valid only at nonzero frequencies is seen to involve the same Green's function.

Representation of operators in terms of incomplete basis sets will cause severe difficulties in maintaining gauge invariance in the theory of magnetic susceptibilties. It is generally required to use field dependent basis functions and to refrain from the use of expansions in terms of c^{-1} at intermediate steps in the development.

Effects of Nuclear Spins

The topics to be covered in this section are concerned with the parameters governing a nuclear magnetic resonance experiment. The chemical shift and indirect spin–spin coupling constants are directly referred to Green's functions. Formulas from the preceding section are used, and the vector potential is taken as

$$\bar{A}(\bar{r})=\tfrac{1}{2}\bar{B}\times\bar{r}+\mathrm{curl}\,\textstyle\sum_g\gamma_g\bar{I}_g/|\bar{r}-\bar{R}_g|, \qquad (13.36)$$

with the notation of Equation (12.46). A spherical average will be performed such that the final result will have the form, [cf. Equation (13.25)],

$$F=F_0-\tfrac{1}{2}\chi B^2+\textstyle\sum_g\gamma_g\sigma_g\bar{I}_g\,.\,\bar{B}+\tfrac{1}{2}\sum_{gh}J_{gh}\bar{I}_g\,.\,\bar{I}_h. \qquad (13.37)$$

We have already dealt with the second term and will now direct the attention to the last two on the right hand side. These contain the *shielding constants* σ_g and *indirect spin–spin coupling constants* J_{gh}. Again we conclude that the $\frac{1}{2}\bar{B}\times\bar{r}$ term contributes only through the $\jmath^0$ containing Green's function.

Shielding constants do not depend upon the choice of origin of the coordinate system, as can be confirmed by considerations similar to those above about general gauge transformations, and we can thus put the origin at the nucleus in question. The formula for the shielding constant at the origin is then, after averaging over all molecular orientations,

$$\sigma=(e/3mc^2)\int d\bar{r}q(\bar{r})/r-(1/3c)\mathrm{Re}\,\textstyle\sum e_{\lambda\mu\nu}\int d\bar{r}\,\langle\langle M_\lambda;\,j^0_\mu(\bar{r})\rangle\rangle_{E=0}(r_\nu/r^3), \quad (13.38)$$

where the antisymmetric tensor $e_{\lambda\mu\nu}$ equals unity for even permutations of (xyz) and minus one for odd permutations.

Indirect spin–spin coupling constants can be identified in a straightforward manner from Equation (13.25), but this gives a rather formidable expression. In order to further elucidate the various contributions we make a few preliminary observations. First it should be noted that the tensor Green's function $\langle\langle \bar{m}(\bar{r}); \bar{m}(\bar{r}')\rangle\rangle_E$ is proportional to the unit tensor when all magnetic fields are absent in the unperturbed hamiltonian H_0. We introduce the notation

$$\mathrm{Re}\,\langle\langle m_\nu(\bar{r}); m_{\nu'}(\bar{r}')\rangle\rangle_{E=0} = \delta_{\nu\nu'} J(\bar{r}, \bar{r}'), \tag{13.39}$$

which will prove to be useful in the further development.

Secondly we observe that the field $\bar{B}(\bar{r})$ is a sum of terms like

$$\begin{aligned}\bar{B}_g(\bar{r}) &= \mathrm{curl\ curl}\,(\gamma_g \bar{I}_g/|\bar{r} - \bar{R}_g|) \\ &= (8\pi\gamma_g/3)\bar{I}_g\delta(\bar{r}_g) + (\gamma_g/r_g^5)\,[3(\bar{I}_g \,.\, \bar{r}_g)\,\bar{r}_g - \bar{I}_g r_g^2],\end{aligned} \tag{13.40}$$

with the notations

$$\bar{r}_g = \bar{r} - \bar{R}_g,\ r_g = |\bar{r}_g|. \tag{13.41}$$

The δ-function contribution to the field gives rise to the Fermi contact interaction while the other term gives the dipole–dipole interaction between nuclear and electronic spins.

Three contributions can be distinguished as separate terms in the indirect spin–spin coupling constants. We take as the first one

$$\begin{aligned}J_{gh}^0 = \gamma_g\gamma_h \Big[&(2e/3mc^2)\int d\bar{r}\,\langle q(\bar{r})\rangle_0\,(\bar{r}_g \,.\, \bar{r}_h)/(r_g r_h)^3 \\ &+ \tfrac{1}{3}\mathrm{Re}\sum \int d\bar{r}\,d\bar{r}'\,e_{\lambda\mu\nu}e_{\lambda\rho\tau}r_{\mu g}\,\langle\langle j_\nu^0(\bar{r});\, j_\tau^0(\bar{r}')\rangle\rangle_{E=0}\,r'_{\rho h}/(r_g r'_h)^3\Big],\end{aligned} \tag{13.42}$$

which arises from orbital contributions to the magnetic moment density in the molecule. The result of spherical averaging is expressed again with the help of the antisymmetric tensor $e_{\lambda\mu\nu}$. The diagonal term J_{gg} appears to be infinite, a difficulty of the theory which is attributed to the assumptions about point dipoles and the nonrelativistic treatment of the electrons. These terms do not influence the interpretation of the nuclear magnetic resonance experiment.

We consider next the contribution from the dipole–dipole interaction term as put forth in the last term on the right hand side of Equation (13.40). The spherical averaging procedure results in the expression

$$J_{gh}^d = \gamma_g\gamma_h c^{-2}\int d\bar{r}\,d\bar{r}'\,J(\bar{r}, \bar{r}')\,[3(\bar{r}_g \,.\, \bar{r}'_h)^2 - (r_g r'_h)^2]/(r_g r'_h)^5, \tag{13.43}$$

which is finite also for g and h equal.

The cross terms between the Fermi contact term and the dipole term of Equation (13.40) vanish when spherically averaged, and we are left to consider the δ-function contributions. They give the simple result

$$J_{gh}^{c} = \gamma_g \gamma_h (8\pi/3c)^2 J(\bar{R}_g, \bar{R}_h). \qquad (13.44)$$

It follows that the total indirect spin–spin coupling constant is given as

$$J_{gh} = J_{gh}^{0} + J_{gh}^{d} + J_{gh}^{c}, \qquad (13.45)$$

and it is the experience that J_{gh}^{c} is the largest term. The other terms are often neglected.

An approximate form of the function $J(\bar{r}, \bar{r}')$ can be determined from the approximation (7.63). The calculation is similar to that of the first section in this chapter and we write the result as

$$J(\bar{r}, \bar{r}') = -(e/m)^2 \sum_{ll'}^{\text{occ}} \sum_{kk'}^{\text{unocc}} u_k(\bar{r}) u_l(\bar{r}) \, [{}^1\mathbf{B} + {}^1\mathbf{C}]^{-1}{}_{(kl)\,(k'l')} u_{k'}(\bar{r}') u_{l'}(\bar{r}'), \qquad (13.46)$$

where the Hartree–Fock molecular orbitals, $u_k(\bar{r})$, are used explicitly. The simple perturbation theory result is recovered when electron interaction terms are discarded in the $\mathbf{B}+\mathbf{C}$ matrix.

Paramagnetic Molecules

It was observed in conjunction with the derivation of Equation (13.6) that special attention needs to be given to the situation where the ground state of the system is degenerate and we wish to consider the low temperature limit. We will examine the case where the ground state has nonzero spin but is spatially nondegenerate, and the system is subjected to magnetic fields as in the previous section of this chapter.

The ground state manifold of states are labelled $|OSM\rangle$ and excited states are similarly labelled $|nS'M'\rangle$. Matrix elements of the perturbation operator V_1 from Equation (13.21) with the vector potential given as in Equation (13.36) are then, within the ground state manifold,

$$\langle OSM|V_1|OSM'\rangle = \int d\bar{r} \, \bar{B}(\bar{r}) . \langle OSM|\bar{m}(\bar{r})|OSM'\rangle. \qquad (13.47)$$

Only this term contributes since we have assumed that there is no spatial degeneracy. Thus we obtain that, at zero temperature,

$$\langle V_1\rangle_0 = (2S+1)^{-1} \sum_M \langle OSM|V_1|\, OSM\rangle = 0,$$

so that our remaining difficulty exists in the Green's function term of Equation (13.6).

The general expression leads to the following formula

$$\lim_{\beta\to\infty} \tfrac{1}{2}\mathrm{Re} \langle\langle V_1; V_1\rangle\rangle_{E=0} = \lim_{\beta\to\infty} \tfrac{1}{2} \Sigma_{nm} |\langle n| V_1 |m\rangle|^2 [(\rho_n - \rho_m)/(E_n - E_m)]$$
$$= -\tfrac{1}{2}\beta(2S+1)^{-1} \Sigma\, |\langle OSM| V_1 |OSM'\rangle|^2$$
$$-(2S+1)^{-1} \sum_{n\neq 0} |\langle OSM| V_1 |nS'M'\rangle|^2/E(nS'), \qquad (13.48)$$

where $E(OS)$ arbitrarily has been taken as the zero of energy. Absence of nuclear spins leads to the Curie law for the susceptibility since then we obtain the free energy as [cf. Equation (13.37)]

$$F = F_0 - \tfrac{1}{2}\beta\,(e^2/3m^2c^2)S(S+1)B^2, \qquad (13.49)$$

where only the divergent term in β has been kept.

Contributions to the matrix elements of Equation (13.47) arise from nuclear spins through the contact field and the dipole field of Equation (13.40). A spherical average will remove the latter contribution and there remains, for one nucleus, the term

$$\lambda \langle OSM\, |V_1^c|\, OSM'\rangle = \langle OSM\, |\bar{I}_g . \bar{S}|\, OSM'\rangle \left[\frac{-8\pi\gamma_g \langle OS\|\bar{m}(\bar{R}_g)\|OS\rangle}{3c\,\langle OS\|\bar{S}|\,OS\rangle}\right]. \qquad (13.50)$$

The factor in brackets is the hyperfine interaction constant a_g in the spin hamiltonian for a paramagnetic molecule, as it is used in the interpretation of paramagnetic resonance experiments. According to the properties of tensor operators we can rewrite the expression as

$$a_g = -[8\pi\gamma_g/3cS(S+1)]\,\langle \bar{S}.\bar{m}\,(\bar{R}_g)\rangle_0, \qquad (13.51)$$

where the average is taken over the ground state manifold of states.

An alternative discussion of the hyperfine interaction constant is given in the next section.

Hyperfine Interaction in States with one Unpaired Electron or Hole

This section is devoted to an examination of the influence of a nuclear spin on the electron propagator. We will consider a closed shell ground state at zero temperature and recall that the poles of $G(\xi, \xi'; E)$ occur at energies corresponding to electron binding energies. Transition amplitudes of $\psi(\xi)$ and $\psi^{\dagger}(\xi')$ occur only for intermediate doublet states in the spectral representation (5.14) and every pole is doubly degenerate. This degeneracy is removed when a nuclear spin is introduced and the magnitude of the splitting between the formerly degenerate poles gives the desired information on the spin density parameter a_g for the corresponding states. This indirect calculation of these constants, which may seem somewhat artificial, has nevertheless a chemical similitude to the experimentally employed methods of ionization or electron attachment for producing paramagnetic species.

For simplicity we limit ourselves to a discussion within the Hartree–Fock approximation. Molecular orbitals, $u_k(\bar{r})$, are assumed to be known for the closed shell ground state in the absence of nuclear spins. The contact interaction from one nuclear spin is added to the hamiltonian and the nuclear spin vector is taken as the z-axis of the coordinate system. The perturbation operator is then

$$\lambda V_1^c = -(8\pi e\gamma_g I_g/3mc)\sum_{kl\nu}\nu u_k(\bar{R}_g)\, u_l(\bar{R}_g)\, \tilde{a}_{k\nu}^\dagger \tilde{a}_{l\nu}. \tag{13.52}$$

It is convenient to introduce the notation

$$d_k = (8\pi e\gamma_g I_g/3mc)^{\frac{1}{2}}\, u_k(\bar{R}_g), \tag{13.53}$$

instead of the orbital amplitude at the nucleus.

Attempting to solve the Hartree–Fock equations to self-consistency we start the first cycle of iteration by taking averages over the unperturbed state. Then we obtain the approximate Fock operator as

$$F_{k\nu,l\nu'} = \langle [\, [\tilde{a}_{k\nu}, H + \lambda V_1^c], \tilde{a}_{l\nu'}^\dagger]_+\rangle_0 = \delta_{\nu\nu'}\, [\epsilon_k \delta_{kl} - \nu d_k d_l]. \tag{13.54}$$

Accordingly we find the electron propagator

$$G_{kl}^\nu\, (E) = \langle\langle \tilde{a}_{k\nu};\, \tilde{a}_{l\nu}^\dagger \rangle\rangle_E, \tag{13.55}$$

from the equation system

$$(E - \epsilon_k)\, G_{kl}^\nu(E) + \nu d_k \sum_m d_m G_{ml}^\nu\, (E) = \delta_{kl}. \tag{13.56}$$

The solution is readily arrived at as

$$G_{kl}^\nu\, (E) = \delta_{kl}/(E - \epsilon_k) - \nu d_k d_l g^\nu(E)/(E - \epsilon_k)\, (E - \epsilon_l), \tag{13.57}$$

where

$$g^\nu\, (E) = [1 + \nu \sum_k d^2/(E - \epsilon_k)]^{-1}. \tag{13.58}$$

There will be corrections to the unperturbed propagator for all states with nonvanishing d_k. These changes will give rise to modifications in the potential terms in the Fock operator. In order to attain the first order corrections to the potential we put $g^\nu(E)$ equal to unity on the Coulson contour and integrate in Equation (5.22) to get from Equation (7.21) that the new Fock operator will be

$$F_{k\nu,l\nu'} = \delta_{\nu\nu'}\, [\epsilon_k \delta_{kl} - \nu d_k d_l - \nu \delta F_{kl}], \tag{13.59}$$

with

$$\delta F_{kl} = \sum_{\kappa\lambda} (k\kappa|\lambda l)\, d_\kappa d_\lambda (\langle \tilde{n}_{\kappa\nu}\rangle - \langle \tilde{n}_{\lambda\nu}\rangle)/(\epsilon_\lambda - \epsilon_\kappa). \tag{13.60}$$

One would normally expect that this perturbation is very small and that it is sufficient to consider the braking of the degeneracy of ϵ_k to first order, that is we observe two levels separated by an amount of energy $\delta\epsilon_k$ such that

$$\delta\epsilon_k = d_k^2 + \delta F_{kk}. \tag{13.61}$$

The spin density parameter for the state k equals

$$a_g = \delta\epsilon_k / I_g. \tag{13.62}$$

The contribution coming from δF_{kk} is called the induced spin density and is of particular importance when d_k is zero, such as it is for states with zero amplitude at the nucleus.

A diagrammatic representation of δF_{kk} comes out to be in the notation of Chapter 11,

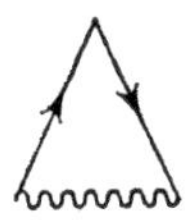

and it is possible to ascertain that the exact spin density parameter should be calculated from the irreducible vertex part.

Problem 1

Consider the Hückel hamiltonian in the presence of a constant homogeneous magnetic field as given by London:

$$H = \sum_{r\nu} \alpha_r n_{r\nu} + \sum_{rs\nu} \beta_{rs} a_{r\nu}^{\dagger} a_{s\nu},$$

where

$$\beta_{rs} = \beta_{rs}^{0} \exp\left[-(ie/2c)\, \bar{B}.(\bar{R}_r \times \bar{R}_s)\right].$$

Identify the matrix elements of the magnetic moment operator $\bar{M}$ of Equation (13.35) from an expansion in powers of B.

Problem 2

Demonstrate that a shift of the origin of the coordinate system in the representation of the hamiltonian in the preceding problem is equivalent to a transformation of the creation and annihilation operators such that

$$\tilde{a}_{r\nu} = a_{r\nu} \exp(i\chi_r),$$

and that this does not influence the anticommutation rules.

Notes and Bibliography

Second order energies related to the calculation of force constants have been discussed by W. Byers Brown (1958), *Proc. Camb. Phil. Soc.* **54,** 251.
The classical reference on electric and magnetic properties of atoms and molecules is the monograph by J. H. van Vleck from 1932, "The Theory of Electric and Magnetic Susceptibilities" (Oxford University Press). A. Carrington and A. D. McLachlan have in their book "Introduction to Magnetic Resonance" (Harper and Row, 1967) an excellent survey of the relevant parameters. The comprehensive volume "Nuclear Moments" (J. Wiley, 1953) by N. F. Ramsey gives thorough analysis of phenomena associated with nuclear spins. The original treatment of the nonlinear field dependence of atomic orbitals in the discussion of molecular susceptibilities was given by F. London in *J. phys. rad.* **8**, 397 (1937).

SUBJECT INDEX